QUANTUM FIELD THEORY
CONFORMAL GROUP THEORY
CONFORMAL FIELD THEORY

by R. Mirman

Group Theory: An Intuitive Approach
(Singapore: World Scientific Publishing Co., 1995)

Group Theoretical Foundations of Quantum Mechanics
(Commack, NY: Nova Science Publishers, Inc., 1995)

Massless Representations of the Poincaré Group
electromagnetism, gravitation, quantum mechanics, geometry
(Commack, NY: Nova Science Publishers, Inc., 1995)

Point Groups, Space Groups, Crystals, Molecules
(Singapore: World Scientific Publishing Co., 1999)

Quantum Mechanics, Quantum Field Theory
geometry, language, logic
(Huntington, NY: Nova Science Publishers, Inc., 2001)

Quantum Field Theory, Conformal Group Theory, Conformal Field Theory: Mathematical and conceptual foundations, physical and geometrical applications (Huntington, NY: Nova Science Publishers, Inc., 2001)

Illustrations enhanced of transversions on the covers are drawn with MAPLE using instructions
on the front cover:

```
z := x+I*y; a := -2+5*I; b := 3+0.5*I; c := 1+3*I; d := 0.3-7*I; w := (a*z + b)*(c*z
+ d)^(-1); r := 2.4;
simplify(w); wr := Re(evalc(simplify(w))); wi := Im(evalc(simplify(w)));
wrp := plot3d(wr,x=2.06..2.08,y=0.73..0.82,color=green,style=patch);
wip := plot3d(wi,x=2.06..2.08,y=0.73..0.82,color=red,axes=none,thickness=3):
display(wrp,wip,orientation=[120,-112]);
```

on the spine, giving the effect of a transversion on the real axis:

```
plot([w,x=0],x=-infinity..infinity,color=[red,green],thickness=3);
```

and on the back cover:

```
z := x + I*h*y; zn := x - I*h*y; a := 1; d := 1; b := 0; c := 0; h := 0.25; w := (a*z +
b)/(c*z + d); wn := (a*zn + b)/(c*zn + d);                           trc :=
simplify(evalc((1/4)*((w + wn)^2 - h*(w - wn)^2)));
implicitplot(trc=1,x=-2..2,y=-10..10,thickness=3, color=red,
grid=[150,150],axes=framed,scaling=constrained);            a := 2 +
4.5*I; an := 2 - 4.5*I; c := -1; d := -10; b := 0.3 + 7*I; bn := 0.3 - 7*I;
w := (a*z + b)/(c*z + d); wn := (an*zn + bn)/(c*zn + d);
trc := simplify(evalc((1/4)*((w + wn)^2 - h*(w - wn)^2)));
implicitplot(numer(trc)=denom(trc),x=-20..20,y=-30..30,thickness=3,
color=red,grid=[150,150], axes=framed,scaling=constrained);
```

QUANTUM FIELD THEORY
CONFORMAL GROUP THEORY
CONFORMAL FIELD THEORY
Mathematical and conceptual foundations
physical and geometrical applications

R. Mirman

AN AUTHORS GUILD BACKINPRINT.COM EDITION

Quantum Field Theory Conformal Group Theory Conformal Field Theory
Mathematical and Conceptual Foundations Physical and Geometrical Applications

AN AUTHORS GUILD BACKINPRINT.COM EDITION

Published by iUniverse, Inc.

For information address:
iUniverse, Inc.
2021 Pine Lake Road, Suite 100
Lincoln, NE 68512
www.iuniverse.com

Originally published by Nova Science Publishers, Inc.

ISBN: 0-595-33692-2

Printed in the United States of America

Preface

P.1 Foundations, geometry, group theory, field theory

Foundations of quantum mechanics and quantum field theory (really one subject) and more broadly of physics are based on geometry and its transformation group, the Poincaré group. These are the topics of this book and also those that it follows and on which it is based [Mirman (1995b,c,2001)]. But geometry leads to a larger group, of which the Poincaré group is a subgroup, the conformal group (which is most relevant for 3+1 dimensions). This then gives the underlying purpose of much of the discussions here, finding the reasons for the relevance of the conformal group to geometry, thus to physics, to understand these connections and the group, to seek clues about how it can give knowledge of our physical universe, and also of many areas of mathematics. The group is unusual and complicated, so providing many interesting suggestions, much insight into physics, mathematics, group theory, but also many, many areas that need be explored, for their own value and to obtain further knowledge and insight about these various subjects. Hence we can only start the process of studying and understanding the group and its implications, in particular to gain guidance about the questions, the problems, the areas whose study should be profitable, and to emphasize the value of these studies with the hope that others will also realize this value and add to our knowledge of these many fields.

Yet while the discussion is merely extended here it does provide interesting results, in particular from the foundations of quantum field theory, such as why baryon (so lepton) number must be conserved, why the proton cannot decay. And there are aspects that suggest possible relationships to experimental regularities that are listed for convenience (appendix B, p. 246).

P.1.a   The unappreciated richness of group theory

Beyond its geometrical, thus physical, relevance, the conformal group is very unlike more familiar ones especially since it is a simple group realized with an inhomogeneous group, the Poincaré group, as a subgroup. Thus it is instructive and should spur readers to see prejudices that come from study of more usual groups, freeing and motivating them to explore group theory, and its applications, in greater depth. (It is not only in quantum mechanics that people have prejudices.)

The conformal group is realized differently (here) than it usually is, with the Poincaré-subgroup-invariants diagonal, a property not of the group, but the realization, so for greater understanding the simplest case of this realization, that for su(1,1), is considered. Seeing these, even if not immediately relevant in other ways, although they may very well be, helps break down predispositions about what realizations and representations are; these are quite distinct. And once thinking has been stimulated, and prejudices made evident, results unforeseen may emerge.

Conformal groups, and related subjects, emphasize how rich group theory really is, something not generally understood. Restricted examples that are well-known tend to bias, so hide richness. Understanding or at least noticing might inspire development of new areas of research, like more general or new types of special functions and solutions of nonlinear differential equations, and of course many areas of physics.

As mentioned several times here group theory plays important roles in many fields of mathematics — and physics. But it seems likely that such roles are restricted because the views of group theory, its realizations and representations, are. One aim is to emphasize these limitations and perhaps inspire work in application of unexplored aspects to, of course physics, but also the many parts of mathematics to which group theory is relevant, and perhaps others to which its relevance has not previously been recognized.

A major aim of the book is thus to stimulate interest in the questions raised by these discussions and detailed study of them, which might prove quite rewarding, perhaps in many ways.

Motivation, rather than comprehensive results, is therefore the more important criterion. Hence the book makes no claim to rigor. The basis of rigor is clarity and comprehensibility. If an argument cannot be widely understood and checked, it is more likely wrong. Obscurantism is the very opposite of rigor. Yet too often mathematical (and even physical) papers and books are written so as to be completely incomprehensible. It is clear that such publications are either vacuous, saying nothing, or just wrong, and are trying to hide their worthlessness under a pile of esoteric, but meaningless, language. While this book makes no attempt at rigor, it does try to avoid the extreme lack of rigor so characteristic of modern mathematics.

P.1.b   The conformal group, geometry and physics

The present discussion covers the conformal group and especially reasons why it is geometrically, so physically, relevant. This again can encourage much thought. The group is interesting in several ways beyond those for which it is currently used, and thus worth further study.

One part of the foundation of modern physics is that of invariance, the laws of physics are the same in all Lorentz frames for example. Neither physics nor geometry can distinguish relatively rotated coordinate systems for space is invariant under rotations. And for the same reason, so it is for relatively boosted frames, these being rotated in an indefinite-metric (3+1) space.

The conformal group introduces other (forms of) transformations, thus other (types of) invariance, (unnoticeable) changes of scale (dilatations), and also those from straight lines to circles, or equilateral hyperbolas, for definite and indefinite metrics. It is clear why straight lines are equivalent, and it is interesting to consider why straight lines are equivalent, in this sense, to circles (or hyperplanes to hyperspheres), or equilateral hyperbolas; what is it about the geometry of (at least locally flat) space that leads to this equivalence? What does it mean? And what is its physical significance? These transformations, in an indefinite-metric space, go from straight lines, paths of motion with constant velocity, to ones that are paths of objects with constant acceleration (although there are also ones going from constant velocity to constant velocity, or constant acceleration to constant acceleration which are not rotations, so add interest). Thus the conformal group makes constant velocity and constant acceleration equivalent. In what sense are these equivalent, since that is not the same as between two paths of constant, but different, velocity (relatively boosted paths)? And what is the physical meaning, what are the physical consequences, of this equivalence? What does it tell about, what does it require of, physics? These are some of the reasons why the conformal group is interesting, enlightening, provocative.

Also the group, according to a long known interpretation, leads to other types of invariance, so suggests reasons for interactions. There might be a fundamental necessity for them. Perhaps this discussion can arouse thinking on such subjects. There are many aspects of the groups that can profitably be thought about, and some are suggested.

There is much to be done and if readers can be inspired to work in these fields there are possibilities of gaining additional understanding of physics, geometry, group theory, special functions, (nonlinear) differential equations and beyond.

Quantum field theory, whose physical foundations have been considered [Mirman (2001)], is here studied mathematically, including the description of multi-particle states and decaying objects (these do not have complex energy of course). This provides a firmer basis for it, allowing development of new and extended results including such as conservation of baryon and lepton numbers. The conformal group, especially with quantum field theory, is also likely to lead to fresh insights about physics, as is discussed.

P.2   ACKNOWLEDGEMENTS

There are many whose comments, criticisms, disagreements, communications, papers, suggestions and other information, even encouragement, over the last few years have stimulated thinking that lead to some of the present commentaries. These include those previously mentioned [Mirman (2001)] and also Luke McAven, Fritz Rohrlich, David J. Rowe, John Ryan, Carl Wulfman, and particularly Frank Sommen. They provided much assistance and guidance that proved so necessary for the thinking that lead to these observations.

And for much help, this is to thank the Department of Natural Sciences of Baruch College (City University of New York).

Table of Contents

Table of Diagrams

Chapter I

Why Conformal Groups are Relevant

I.1 Conformal invariance of geometry and physics

It is clear that much, perhaps all, of physics is determined by geometry (especially?) through its transformation group, the Poincaré group: that nature is quantum mechanical, the dimension of space, the properties of electromagnetism and gravitation, and so on [Mirman (1995b,c; 2001)]. (Of course group theory is also a major determinant of structures and behavior of materials and objects of which they are composed [Mirman (1999)].) And that physics takes place, can take place, in the geometry places potent constraints on it. However there is a larger group of which the Poincaré group is a subgroup, the conformal group, that also is determined by, and expresses, fundamental characteristics of our geometry, for reasons we here consider. Does this group as well have physical significance, beyond that of its subgroup?

It is larger, more complicated, and relevant representations are of unfamiliar kinds as we see, so detailed study is a vast undertaking. But it is important to at least begin this, to outline reasons for its geometrical and physical relevance, to indicate why the structure of its suitable representations is different from more familiar ones, and to begin to see what must be learned. We try to start this here, certainly with the hope of encouraging more extensive explorations.

One purpose of these analyses is to understand the conformal transformations of geometry and their implications. In particular since there is no intrinsic length geometry is invariant under the similitude group, the Poincaré group plus the dilatation transformations — a conformal subgroup. If we carry out such a transformation there is no way of distinguishing the geometries before and after — all properties are the same. Undoubtedly this has physical application. The meaning of the remaining generators, the transversions, is perhaps more subtle, although there are interesting hints (sec. I.7, p. 55).

How does this discussion relate to the previous discussion [Mirman (2001)] on quantum field theory and its interpretation, and also to that below (chap. IV, p. 166)? The fundamental view here as before [Mirman (1995b,c)] is that physics is determined (completely?) by geometry, and that geometry is limited (determined?) by the ability of physics to take place within it. Much of the discussions, here and elsewhere, are based on this. Thus it is reasonable to try to extend the analysis to see whether geometry can lead to even greater understanding of physics. While we are able to only begin, it is clear that with this approach — basing physics and geometry on each other, through group theory — all these topics, though they may seem at first very different, are closely related.

I.1.a   The relevance of groups even without invariance

Physics may not seem invariant under dilatations — there are lengths given by masses of particles, Compton wavelengths. But even if this dilatation invariance did not hold, which it actually does, a reason that the conformal group would still be relevant can be seen from the analogy with the rotation group and particles in a magnetic field for which physical laws are not invariant under rotations (if the magnetic field is not rotated). Nevertheless statefunctions are rotation-group basis states; they have to be since the rotation transformations must be able to transform them. Moreover rotations take states of one representation into sums of states of the same representation, whether physics is invariant under them or not. This is a consequence of properties of geometry whose transformations include rotations. Invariance of physics is irrelevant to the set of possible statefunctions, and that states mixed by a rotation all belong to the same representation [Mirman (2001), sec. I.7.b, p. 37; sec. I.8, p. 44]. (Statefunction is a better term than wavefunction [Mirman (2001), sec. IV.1.a, p. 143].)

It is the same for the conformal group. The most general transformations of the geometry are those of it. Thus physical statefunctions must be its basis vectors. That a rotation acting on statefunctions takes them into ones with different properties (different energies or lifetimes, say) does not eliminate the requirement for statefunctions to be rotation basis states. Likewise that a conformal transformation takes a statefunction into one with different properties (say with a mass no longer going with a particle, which it does not) cannot remove the (geometrical) requirement that it be a conformal group basis state. However the reasons are stronger, and there is dilatation invariance even though particles have mass.

I.1.b What is transformed?

When we say that physics is invariant under a set of transformations, what transformations are we considering? Some object is transformed relative to something else; what is it transformed with respect to?

That physics is rotationally invariant has meaning as can be seen by considering whether it cannot be. It is possible to think of spaces whose properties vary with direction. Thus if space were the shape of a cylinder or a torus, there would be clear distinctions between directions. Then geometry, hence physics, would not be invariant under all rotations. Rules for geometry, thus laws of physics, would be directionally dependent. Consider a space shaped like a pretzel. While we might take a tangent plane, and consider rotationally invariance on that, geometry, so physics, would vary greatly with any slight displacement, and this would depend on direction (especially if the pretzel were salted). For these rotational and translational invariance do not hold. So when we say that space and physics are invariant, this has definite content, as we can easily find geometries and laws of physics for which invariance does not hold. There are shapes, cylinders, pretzels, and so on that do not allow some invariances, but also spaces such as ones consisting of points of a crystal (really lattice) that are invariant not under the (inhomogeneous) rotation group but only under a subgroup [Mirman (1999), sec. I.3.a, p. 6].

Space is invariant, not because the laws of physics or geometry cannot tell whether one object is transformed with respect to another — of course they can — but because it is impossible to tell if an entire system has been Poincaré transformed at all.

Conformal invariance then also has meaning, for it might not hold, as for example if there were an intrinsic length in space, as is true for a crystal. Requiring it places restrictions on physical laws.

And this is a property of geometry. It does not imply a transformation of one object with respect to another, but rather an inability to tell whether a transformation has been carried out, just as with rotations we could not tell if one was performed if an entire system is rotated (unlike counter-examples just mentioned).

I.1.c The reason for the appearance of the rest mass

Why is the rest mass used (sec. I.3.b, p. 24), explicitly or implicitly throughout — the mass value is independent of velocity and acceleration? Reasons for it are fundamental — to how we must (?) manage our understanding of physics — and are part of the foundation of this discussion (and probably of all such discussions).

It follows from the relativistic version of Newton's second law [Anderson (1967), p. 197], since force is independent of mass. A better argument is based on Dirac's equation (and Newton's second law is the classical, nonrelativistic limit of Dirac's equation [Mirman (1995b), sec. 6.3.3, p. 118]) with an interaction term giving acceleration. The mass in it does not depend on the interaction term. This equation, for spin-$\frac{1}{2}$, is equivalent to the two Poincaré invariants [Mirman (1995b), sec. 6.3, p. 114], and we assume that nonlinear terms are inserted in the equations for these invariants (specifically the one for mass), with the value of the invariant, the mass, unchanged.

Actually for any object there are equations, for spin-$\frac{1}{2}$ only one is needed, which are its properties. Thus we can write Dirac's equation for say an electron in general; the mass is that experimental value found for an electron not in a field. Placing it in a field determines the particular expression for the interaction. But the form of the general equation itself is independent of this particular value. If that were not true, then equations governing objects would depend on specific circumstances, and physical laws would be (at least close to) impossible.

The equation for the Poincaré mass-invariant is actually an operator equation (acting on basis vectors). For a free particle it can be written

$$\frac{d^2}{dt^2} - \sum_i \frac{d^2}{dx_i^2} = m_o^2; \qquad (I.1.c\text{-}1)$$

applied to a basis vector it picks out those belonging to representations labeled by m_o. With interactions it becomes

$$(\frac{d}{dt} + H_{int})^2 - \sum_i (\frac{d}{dx_i} + P_{i,int})^2 = m_o^2, \qquad (I.1.c\text{-}2)$$

where the added terms are schematic symbols for the proper nonlinear expressions that are relevant for the particular object. For an electron they would be the electromagnetic, weak and gravitational interactions.

It is a fundamental assumption (not only here) that interactions are included by adding them to momentum operators. This must be true for inclusion of massless fields, as is well-known and well-understood [Mirman (1995c)], but there seems no proof that it is the correct way for strong and weak interactions. Yet it is (at least) reasonable to do that. The definition of an object, an electron say, is independent of its acceleration (sec. I.4.c.iii, p. 41). If interactions were not put in this way it is unlikely they could be inserted in any way, at least in a way that allows general laws.

With interactions states are basis states of nonlinear representations, and very little is known about these — the number of questions is vast. Perhaps there are nonlinear representations (or realizations) for which nonlinear terms do not appear in this way, that is for which momentum operators are not of the form

$$P_\mu = i\frac{d}{dx_\mu} + P_{\mu,int}. \qquad (I.1.c\text{-}3)$$

Even if they should exist but not have physical applications, they might have mathematical ones. Nonlinear representations provide much to explore.

It is always (everywhere) assumed that interactions are included by adjoining them to momentum operators, and in this form.

I.1.d   Who makes measurements

In these discussions we compare measurements made by different observers, some moving with constant velocity, others perhaps accelerated. Which observers do we use, how do we, and they, know that they are the ones specified, and why do we choose them? We have to specify which measurements each makes, how they are able to do so, and how they can compare what they find — and whether they can interpret their measurements so as to give information about physical laws.

Of course, the most important observers are inertial ones, assuming that we can find them, and that they themselves can tell whether they are — for it is the observations that they report that give us physical laws. And we cannot correct observations by noninertial ones unless we are ourselves inertial (or can refer our observations to inertial frames), and can know that we are.

I.1.d.i   *Geometrical and physical foundations of straight lines and constant velocity*

How does an observer know that it is moving with constant velocity? Geometrically a straight line has a well-defined meaning — straight lines are intrinsic properties of (pseudo-)Euclidean geometry. However, as we see (sec. I.5, p. 43), it is possible to distort the geometry but leave straight lines straight, for it can be distorted so that it is impossible within it to tell that it has been transformed — straight lines are involved in the definition of the geometry so not all changes in them are detectable.

Physically it is possible to look at an object, a hydrogen atom for example, a completely well-defined object, independent of its, or the observer's, motion, so with fully-known states. An observer can specifically look at an atom at rest with respect to itself, determining its energy levels and transitions between them. The laws of physics are the same in all uniformly moving systems so experiments like these find which systems are uniformly moving. But an atom in a field, so accelerated, has its energy levels, and other properties, affected.

In another experiment an observer looks at an atom (really a collection) moving with respect to it. The velocity of the atom can be established when it coincides with the observer, so only purely local measurements are needed. Then radiation it emits can be measured at different times, and for an observer and atom both moving with constant velocities, its spectrum, the frequencies of emitted radiation, are determined by the (known) spectrum of the atom and the Doppler shift. Since the velocity is known (or found this way), this tells whether the velocities of the two systems are constant. If so, but only if they are, the Doppler shift is constant in time. Thus an observer can check that it is moving uniformly (by finding collections of atoms with constant Doppler shift) — it can verify that it is an inertial observer. Only such observers see radiation given by known energy levels of objects at rest, but with constant Doppler shifts.

For an observer uniformly accelerated along z, the curvature of its path in the z, t plane can be found, a geometrically well-defined and meaningful value. It can look at a system with constant velocity, which can inform it that it is so moving (its energy levels tell that), and by comparing its velocity at different times (say by using the Doppler shift) can find its own acceleration, and thus its path in the z, t plane. There is a geometrical difference between this path and one for constant velocity, a straight line, so an invariant physical difference between constant velocity (including zero) and nonconstant velocity, acceleration. There is also, as we discuss, a geometrical, so physical, difference between constant and nonconstant acceleration. And there are geometrical transformations between paths of inertial objects, straight lines, and those of uniformly accelerated ones, equilateral hyperbolas (sec. II.3.d, p. 82).

I.1.d.ii *Physical laws seen by an observer with constant acceleration*

How do laws of physics seen by a uniformly accelerated observer compare with those seen by observers with constant velocity? At each instant, such an accelerated observer can compare its observations with those of an inertial system, instantaneously at rest with respect to, and coinciding with, it at that instant. It can check that the path of the latter is tangent to its own, by seeing that they do not cross at any time. By comparing observations at various times it can measure its own acceleration, and verify that it is nonzero and constant.

We study measurements by inertial observers on objects both with constant velocity and with constant acceleration, and compare them to find how properties of such objects are related; how does acceleration affect properties such observers measure? We do not compare observations of an object with constant velocity made by inertial and by accelerated observers. These give different questions, more difficult to study, and it is useful to note why.

Consider observations by an observer (with zero velocity) on an object and on a rotated one, and also on one object by an observer and by a rotated one. These give the same results, whether the object or observer is rotated. Since all such observers, and all such objects, are equivalent — their physical properties are identical — only angles between observers and objects can affect observations.

However inertial and accelerated observers are not identical, they differ physically in significant ways. Thus we cannot directly determine observations of an accelerated observer from those of an inertial one.

To clarify this we have to study the meaning of measurement.

I.1.d.iii *What is the fundamental definition of measurement?*

What is a measurement? On a fundamental level it is a determination of the final statefunction of an observer from its initial one. Take an observer governed by equations with interaction terms containing the statefunction of the object. We want a measurement by an inertial observer — actually an oxymoron since if an observer makes a measurement, the object acts on, so accelerates, it; however we assume that it was inertial before the measurement, and the affect on it is "small".

So we consider the statefunction of the system to be that of the observer, a wavepacket, times that of the object, with (almost) no overlap before a certain time, and with (almost) no overlap after some time later. The final statefunction of the observer gives the measurement; this is determined by properties of the object (such as mass and spin) — these can be extracted from the statefunction. We discuss here only such measurements, first for a statefunction of an object with constant velocity, then the same object with constant acceleration. Information about the object is then extracted from final statefunctions of the observer and compared.

Of course doing this would be difficult. However we assume that properties of the object so found are those given by the mathematical transformations of the group acting on its statefunction. If not physics would be inconsistent. This assumption of the identity of properties found from measurement with those given by mathematical transformations forms part of the foundation of the study of transformations, symmetry, and everything based on them.

I.1.d.iv *Why accelerated observers are not the same as inertial ones*

We can do the same for an accelerated observer (in principle). However instead of taking its statefunction to be a wavepacket of plane wave states, we take it as a wavepacket of accelerated statefunctions. We then find the properties of the object and compare them with those found by an inertial observer. However it is unclear that the accelerated observer is able to extract properties that can be correlated with those found by inertial observers. It is not clear that an accelerated observer can in any meaningful way assign the object mass, spin, Poincaré representation and so on.

Further there seems no reasonable way of transforming properties that an inertial observer measures to those found by an accelerated one, as those of rotated observers can be, and as we can relate properties of accelerated and non-accelerated objects. Mathematical transformations, at least of the form considered here, correlate statefunctions of objects and we can ask how properties of these correspond? But for different observers their statefunctions do not (in any simple way) provide information about their measurements.

It might be possible to study the physical processes that form measurements and compare the two types of observers. But then there would be more questions about whether results are the same as rigorously found from final statefunctions. It would be a serious problem if that were not true for inertial observers. But observations by accelerated observers have no immediate fundamental relationship to physical laws.

Thus we limit considerations to inertial observers, and just raise these questions for others, with the possibility that studying them could be useful.

I.1.e The conformal group

While the conformal group is studied in depth below (chap. III, p. 106), we list here its transformations to define terms. It has first the transformations of its subgroup, the Poincaré group, with Lie algebra generators

$$M_{\mu\nu} = -M_{\nu\mu}, \quad \mu, \nu = 1, 2, 3, 4 \quad (\dots \text{for higher-dimensional spaces}),$$
$$(\text{I.1.e-1})$$

the homogeneous generators, and momentum operators P_μ, the inhomogeneous ones. Generators M_{ij}, $i, j = 1,2,3$, are rotations, M_{4i} pure Lorentz transformations, boosts. In addition to these is the dilatation,

$$x'_\mu = \rho x_\mu, \qquad (\text{I.1.e-2})$$

with generator D. Most interesting are special conformal transformations — transversions (eq. III.1.c-3—III.1.c-4, p. 108; eq. III.1.d.vi-20, p. 114),

$$x'_\mu = \sigma(x)^{-1}(x_\mu + c_\mu x^2), \qquad (\text{I.1.e-3})$$

$$\sigma(x) = 1 + 2c_\mu x_\mu + c^2 x^2, \qquad (\text{I.1.e-4})$$

with generators K_μ. These are the fifteen generators of the group (in 3+1 space), 10 from the Poincaré group, the dilatation, and the four transversions.

The conformal-algebra generators obey the commutation relations (sec. III.1.d.v, p. 112)

$$[D, P_\mu] = iP_\mu, \quad [D, K_\mu] = -iK_\mu, \quad [D, M_{\mu\nu}] = 0, \qquad (\text{I.1.e-5})$$

$$[K_\mu, K_\nu] = 0, \quad [P_\mu, P_\nu] = 0, \qquad\qquad \text{(I.1.e-6)}$$

$$[K_\mu, P_\nu] = 2i(g_{\mu\nu}D - M_{\mu\nu}), \qquad\qquad \text{(I.1.e-7)}$$

where $g_{\mu\nu}$ is the metric,

$$[K_\lambda, M_{\mu\nu}] = i(g_{\lambda\mu}K_\nu - g_{\lambda\nu}K_\mu), \qquad\qquad \text{(I.1.e-8)}$$

$$[P_\lambda, M_{\mu\nu}] = i(g_{\lambda\mu}P_\nu - g_{\lambda\nu}P_\mu), \qquad\qquad \text{(I.1.e-9)}$$

$$[M_{\kappa\lambda}, M_{\mu\nu}] = i(g_{\lambda\mu}M_{\kappa\nu} - g_{\kappa\mu}M_{\lambda\nu} - g_{\lambda\nu}M_{\kappa\mu} + g_{\kappa\nu}M_{\lambda\mu}). \qquad \text{(I.1.e-10)}$$

I.2   Why the conformal group is fundamental

This discussion is based on the requirement that laws of physics be in accord with the conformal group, that statefunctions be conformal group basis vectors. Why should this be so? What does it mean? And what are its implications?

The Poincaré group (a conformal subgroup) is relevant (to geometry, thus to physics) because it is the transformation group of geometry [Mirman (2001), sec. I.7.b, p. 37]. Thus for any observer there is another (possible) one related by a Poincaré transformation. All these observe (in principle) all physical phenomena, and geometry requires that their observations be related by Poincaré transformations, whether space, or physical laws, are invariant under them. Statefunctions then must be basis states of representations of the Poincaré group — so that they can be transformed by it [Mirman (2001), sec. I.8.a, p. 44]. Is this also true for the conformal group?

Besides Poincaré transformations there are two other sets in the conformal group, transversions and dilatations. We start with the latter.

I.2.a   Physics is dilatationally invariant

It is often believed that conformal invariance does not hold with massive objects present, since its transformations (most noticeable dilatations) change mass. One argument is that while a rotation takes a physically possible state into another physically possible state, although perhaps with different properties, a dilatation acting on the mass of a particle takes the physical state into another that is not possible. Is this noninvariance really possible?

I.2.a.i   *Masses have no meaning, only ratios do*

Mass has no significance — it cannot be defined or measured. Only ratios of masses have meaning. This we often forget because an arbitrary standard mass is defined, the ratios of all other masses with respect to it are found, and masses of objects are then equal to their ratios since the standard mass has been assigned value 1. This tends to hide that these masses are actually ratios.

Since there are no lengths in the geometry, it is not correct to have these transformations act on masses. What is fixed, independent of transformations, are mass ratios, the ratio of the mass of a proton to that of an electron, for example. But masses themselves do not signify anything, they are unmeasurable. Thus all masses can be changed without any effect on physics — there is no way of telling whether they have been. It is only possible to tell if ratios are changed, and they are not by these transformations (of geometry, which change all masses the same way, and corresponding lengths the inverse way (sec. I.2.a.viii, p. 14), leaving ratios fixed).

A dilatation multiplies all masses at once by the same factor, leaving all ratios invariant — all must be changed together. This is similar to a spinning body in a magnetic field (say produced by another spinning object). If only the spin, or only the field, is rotated then the physical situation is changed. Rotational invariance holds — but only if both are identically rotated simultaneously. The same is true for dilatation invariance, all masses must be multiplied by the same factor, and all distances by the inverse one (sec. I.2.a.v, p. 12).

Rotations around the z axis can change phases. That rotations give different statefunctions does not mean that space is not invariant since an overall phase has no physical content, it cannot be measured. Only phase differences (and ratios of masses and lengths) are measurable, and these are invariant under rotations — thus physics is.

Statefunctions then are basis states of the similitude group. They must be since these are transformations of the geometry, and such transformations act on statefunctions also. There must be a group of transformations of statefunctions that is the same as (or at least homomorphic to) the group acting on the geometry — these result in the transformations of the former. (Statefunction is a better term than wavefunction [Mirman (2001)]).

I.2.a.ii *Statefunctions transform under the rotation subgroup, and the entire similitude group*

Statefunctions transform under the rotation group for example since there are differently rotated observers who look at an object, and their observations must be related, and — because of the geometry — they are related by the rotation group (and, of course, the Poincaré group). Statefunctions must thus be basis states of it so that they can be so related. They must also be basis states of the similitude group (the Poincaré group plus the dilatations) because — as there are no lengths in our geometry — there are different observers who choose lengths differently, their observations must be related, and the relationship is given by these transformations of lengths, by dilatations and other transformations of the similitude group. Statefunctions hence must be transformable by it, that is be basis states of the group.

We cannot prevent one observer from choosing a unit length differ-ent from that of another (as is quite clear from the multitude of unit lengths in common use) because there is no fundamental physical way of preferring one unit over another (although some may be more conve-nient because of sizes of our bodies or our instruments, but this does not place restrictions on physical laws). So laws of physics must be such as to not distinguish different unit lengths (and therefore masses) — they must be dilatationally invariant.

Thus while it is believed that dilatation invariance is only approxi-mate because particles have mass, it must be — and is — exact.

I.2.a.iii *Why all terms in an equation have the same units*

Why are dilatation transformations relevant? If all masses in the uni-verse were doubled, and all lengths halved, it would not be possible to tell, there would be no experimental difference, no way that we could determine that such transformations actually took place. Thus such a statement is senseless, for just that reason. But for just that reason physical laws must be invariant under these transformations else they would allow us to observe effects that we cannot. We cannot because there is nothing in geometry that sets a scale, absolute lengths and masses therefore have no meaning, they are nonexistent.

It is often assumed that there are restrictions on physical laws be-cause all terms in an equation have to have the same dimensions. How-ever this is misleading. If we pick a unit of mass — always [Mirman (2001), sec. V.4.f.iii, p. 234] with $h = c = 1$ [Mirman (1995b), sec. A.1.1, p. 177] — say the mass of the electron, all coupling constants (and all other quantities) are dimensionless. Then we cannot correctly argue using a requirement that the dimensions of all terms of an equation must be the same — there are no dimensions. It is necessary to use dilatation invariance, and this places conditions on interactions, cou-pling constants, and other parts of a theory, requiring that all terms in an equation transform identically under these transformations. This is the actual reason for the common belief that all units must be the same.

I.2.a.iv *Dilatations and gravitational mass*

Gravitation has an intrinsic mass, the Planck mass, and since masses are not meaningful, only their ratios, this mass is really a ratio with respect to the unit (sec. IV.3.e, p. 209). That is, if the unit mass is multiplied by λ, this mass must also be, so for the gravitational constant

$$G \Rightarrow \frac{G}{\lambda^2}. \qquad \text{(I.2.a.iv–1)}$$

Thus coupling constants cannot be measured either (except for those that are pure numbers), but only the proper ratios with respect to the units of the system chosen. This is also true for weak interactions, for which the (nonfundamental) Fermi coupling constant has a built-in mass (sec. IV.3.d.ii, p. 207). That there are intrinsic units in coupling constants (but not geometry) has interesting consequences as we see (sec. IV.3.c, p. 201).

I.2.a.v *Conditions are placed by dilatation invariance on units*

It is generally believed that there are three basic units, c, h and a mass. However if we took the unit of length in the x direction to be different from that along y, there would be another universal constant, the ratio of these units. It is questionable whether that would lead to additional conditions on physics, or to additional understanding. But c, the speed of light, is the ratio of the unit of distance in the space direction to that in the time, and should be taken as 1 for the same reason that the units for the x and y directions are taken the same (except for purely calculational convenience, and there might be cases for which different units are useful for the, say vertical and horizontal directions, or in crystals [Mirman (1999)]).

This then gives two units, of length and of mass (or energy, or momentum). But these are dual to each other, units of x and p are inverses since

$$p = i\frac{d}{dx} \quad \text{(giving } exp(ipx)\text{).} \tag{I.2.a.v-1}$$

(Momenta and coordinates are vectors, so should have indices. But we usually treat them schematically and suppress indices. In general details that are unneeded for this explanatory discussion are suppressed for they might otherwise distract from concepts that should be emphasized.) In $exp(ipx)$ the value for which

$$px = 2\pi \tag{I.2.a.v-2}$$

is independent of the units. Thus if we change the unit of x we must change that of p. We cannot change the scale of distance coordinates without also changing the scale of momentum coordinates, thus that of masses (since mass is given by

$$m^2 = p^2). \tag{I.2.a.v-3}$$

A picturesque way of seeing this is to consider $sin(px)$ and a stick, one end at the origin, another point (say the other end) coinciding with the value of x for which

$$px = 2\pi, \quad sin(px) = 1. \tag{I.2.a.v-4}$$

Suppose that all lengths in the universe were multiplied by a factor λ. The coincidence of the maximum of the sine wave and the point on the stick would be unchanged. Physically it would be impossible to tell that all lengths were revised. But this can only be if

$$p \Rightarrow \frac{p}{\lambda}. \qquad \text{(I.2.a.v–5)}$$

Thus distance and momentum (so mass) must vary inversely.

Different units for x and p, as for x and y, give an arbitrary constant, a unit conversion factor (and then we have to carry a physically meaningless constant, which cannot give conditions on physics). If we change the units of x, but not of p, we change the conversion factor. However dilatation invariance gives restrictions on laws of physics, and we cannot evade these by arbitrarily picking different units for x and p and introducing constants (just as we cannot hide rotational symmetry, or evade its consequences, by using different units for x and y). All that such a constant does is confuse and hide what is being done, what requirements are, and how they arise. Here we are interested in conditions placed on nature by conformal invariance of geometry, so set all arbitrary constants to 1 (or perhaps other convenient values).

The unit of distance determines the unit of its dual, momentum, and conversely. This leaves one arbitrary unit to be picked, length, mass, momentum, energy, whichever is convenient. Mass seems the most reasonable.

I.2.a.vi *Mass and distance are meaningless; only ratios have meaning*

We have then one unit, mass or distance. But mass, and distance, are meaningless, empty terms, they cannot be determined. Only ratios are measurable, so meaningful (sec. IV.3.c.ii, p. 203). And ratios are independent of units. Thus if we choose $c = h = 1$, and some unit for mass (like that of the electron), there are no units to give the requirement that all terms in an equation must have the same units. Restrictions regarded as coming from this condition really come from dilatation invariance — from the equivalence of all observers, no matter what unit they choose for mass (or distance). Dilatation invariance then has fundamental physical significance, and imposes restrictions on physical laws. And this is often hidden, and misunderstood, because arbitrary constants are introduced into equations.

I.2.a.vii *Dilatation invariance of geometry, so of physics*

Thus as with Poincaré-group transformations, for any observer, there is another (possible) one who chooses the unit of length (or mass) to have some other value, any other value. But these can have no physical or geometrical significance — there is nothing in the geometry that determines them, there is no way that they can be determined physically (experimentally). Observers are all equivalent — physical laws must not, and cannot, distinguish them. While their units are different, their ratios for physical quantities are the same — invariant under a change of units. This shows why and how the similitude group imposes restrictions on physics.

Physics is of course not invariant under dilatation of only part of a system, just as it is not under rotations of part of a system. But it is invariant under rotations of entire systems, and dilatations of entire systems.

So the fundamental reason for dilatation invariance is, like the reasons for the Poincaré group and its subgroups, geometrical: there is no length in the geometry, physical laws are thus invariant under changes of the unit of length, so mass. This indicates, again, how geometry imposes its will on physics.

I.2.a.viii *The Dirac and Klein-Gordon Equations*

Relationships between mass, or momentum, and coordinates, are also shown, when unnecessary arbitrary constants are not used, by Dirac's equation,

$$i\gamma_\mu \frac{d\psi}{dx_\mu} + m\psi = 0, \tag{I.2.a.viii-1}$$

and the Klein-Gordon equation,

$$(\frac{d^2}{dx_\mu^2} - m^2)\psi = 0. \tag{I.2.a.viii-2}$$

Of course if there is more than one mass, the unit of coordinates must be the inverse of that of one of them while other masses are expressed in terms of that mass unit. Then these equations, for the other masses, will have constants, the ratios of the masses to that of the unit — such ratios have physical content, while values of masses do not.

Dirac's equation goes to, under a dilatation,

$$(i\gamma_\mu \frac{d}{dx_\mu'} + \lambda m)\psi' = 0, \tag{I.2.a.viii-3}$$

with a different (but unmeasurable) mass. However for form invariance,

$$\frac{d}{dx_\mu} \Rightarrow \frac{d}{d\lambda^{-1}x_\mu}, \tag{I.2.a.viii-4}$$

showing again that mass and distance are dual, so vary inversely.

With electromagnetic interactions the equation becomes

$$(i\gamma_\mu \frac{d}{dx_\mu} + m + \gamma_\mu A_\mu)\psi = 0, \tag{I.2.a.viii-5}$$

and is transformed to

$$(iy_\mu \frac{d}{d\lambda^{-1}x_\mu} + \lambda m + \lambda y_\mu A_\mu)\psi = 0, \qquad \text{(I.2.a.viii-6)}$$

which changes A. But A has units of momentum so the equation is invariant.

Thus dilatation transformations leave all equations of the theory invariant — any changes are unphysical, they cannot be detected experimentally. Fields then, such as ψ, the statefunctions, are representation basis vectors of the 3+1-dimensional similitude group (sec. IV.3.c.i, p. 202).

I.2.b   Transversion transformations and geometry

The conformal group is physically important because it is geometrically important. We now understand the similitude subgroup, and that geometry is invariant under it because it is invariant under the Poincaré subgroup and in addition there are no lengths in the geometry; why are transversions relevant? The reasons are based on certain curves of geometry being special. So this we consider next.

I.2.b.i   *The special transformations of geometry*

Transversions can be interpreted as giving constant accelerations, or instead as position-dependent dilatations (sec. I.5, p. 43). Physics is (interestingly) invariant under the group of transformations of geometry, rotations (giving transformations to all differently-oriented coordinate systems), boosts (giving transformations to all frames moving with constant velocity — along straight lines, so that boosts are rotations in non-positive-definite metric spaces), and translations (giving transformations to all different points of space) — physical laws are invariant under the Poincaré group, its subgroups, the translation and Lorentz groups, and the subgroup of the latter, the rotation group. Thus the set of points of space, the set of orientations of coordinate systems and the set of coordinate systems moving with constant velocity are special, both in geometry and in physics. Indeed the points and coordinate system define the (local) geometry — but these transformations and the points and coordinate system they give do not depend on invariance, being purely determined by, and determining, the local geometry [Mirman (2001), sec. I.7.b, p. 37].

These, the relationship of geometry and its transformation group, plus the additional property of the invariance of space under this group, are perhaps the most fundamental facts of nature, so determining the foundations of physical theories, and their correctness. To this list we add the set of frames with different units of length, so mass (related by dilatation transformations), and also the set of frames moving with constant acceleration (geometrically along equilateral hyperbolas), but can add nothing else. Why geometrically and physically is constant acceleration special, and why must we stop with that? To study this we have to see how the conformal group is related to geometry, what curves it gives and what their geometrical significance is.

I.2.b.ii   *Accelerations and boosts*

The arguments here are for the importance of the conformal group, the group whose transformations leave invariant the set of paths of constant velocity, and of constant acceleration [Hill (1945,1947,1951)]; most fundamentally the group that leaves angles fixed. Why stop there? Why not consider larger groups, one (perhaps) leaving invariant sets of paths of nonconstant acceleration? Why is the conformal group special? To examine this we have to first show that transversions do transform to paths of constant acceleration, to find why this is true, its meaning, and what it says about the group, the physical requirements it imposes, and why they should hold. Constant acceleration has geometrical significance because paths of objects so accelerated, equilateral hyperbolas (in non-definite-metric spaces (sec. I.3.b, p. 24)), join straight lines (and circles for a definite metric space) in defining the geometry. This suggests that they should have physical significance also — what this is has to be determined, as does how connections of the groups to geometry and to physics are related.

There is another interpretation of these special conformal transformations (sec. I.5, p. 43), but we first discuss them in terms of constant accelerations to provide one physical terminology going with the geometrical curves.

I.2.b.iii   *Invariance of paths for constant acceleration*

For any constant acceleration there is a path (an equilateral hyperbola) that is invariant; all points are identical. An equilateral hyperbola is one obtained from a circle by making one coordinate imaginary — all such hyperbolas are identical, just as circles are, except for size, given by the radius of the circle.

Transversions leave invariant the set of these hyperbolas — paths of objects with constant acceleration (including zero, which includes velocity zero) — just as boosts leave the set of paths of constant velocity, straight lines, unchanged, and rotations the set of differently oriented frames at rest with respect to each other. Geometry then regards straight lines, circles (in definite-metric spaces) and equilateral hyperbolas (in indefinite-metric spaces) as special.

I.2.c Why the conformal group is related to geometry

The conformal group has transformations that leave invariant the set of straight lines and circles in definite-metric spaces, in indefinite-metric spaces equilateral (rectangular) hyperbolas, and their generalizations, planes, hyperspheres, and hyperboloids in higher-dimensional spaces. Transformations of the conformal group of a space takes these subspaces into the same set of subspaces (sec. II.3, p. 77). Straight lines are special cases of circles (or equilateral hyperbolas), planes of spheres (and similarly for other spaces), those with infinite radii. Thus there are conformal transformations taking lines to circles, and (since the conformal group is a group) conversely.

A transformation of the conformal group takes one point on a circle (or equilateral hyperbola for a space with non-definite metric), including the special case of straight lines (generally hyperplanes), to another, translates the curve (giving the Poincaré subgroup), changes it to an equivalent curve with different radius (giving the similitude subgroup), and for the transversions, takes one circle (or equilateral hyperbola, or corresponding hyperspheres or hyperboloids) to another. A dilatation multiplies the radii of all circles by the same amount — it shrinks or expands space uniformly. But a transversion shrinks space differently at different points. Thus while a circle (or equilateral hyperbola) with finite radius goes under a dilatation to another circle with finite radius, although a different one, under a transversion it can go to one with infinite radius, and conversely. This we see below (chap. II, p. 66) for a two-dimensional space (the complex plane).

These lines, planes, spheres, equilateral hyperbolas, and generalizations, are special, deeply related to geometry — the fundamental curves of the geometry. Why? They are the curves generated by the operations of the geometry's transformation group, the Poincaré group, or its generalization to other dimensional spaces. There is a one-to-one relationship between points on these curves and geometrical transformations. Points of space, and coordinate systems, are equivalent to transformations of the group, and thus these special curves are equivalent to, and define, the geometry. It is a subgroup of the Poincaré group, the rotation group, whose transformations produce circles — for indefinite-metric spaces the Lorentz group, which gives (also) equilateral hyperbolas — but no other curves.

I.2.c.i *There are curves that define, and are defined by, a geometry*

The Poincaré group is the Lorentz group plus the translations, so the geometrical relationships of circles to rotations (points on a circle, or sphere, correspond to rotations), and likewise of equilateral hyperbolas to pure Lorentz transformations, indicate why special conformal transformations, transversions, have distinctive significance for geometry.

Transformations of the conformal group, acting on a point, such as the origin, or an axis, sweep out these curves, the circles (or spheres or hyperspheres and their generalizations to non-definite metric spaces). Curves (swept out by transversions) are related by a series of Poincaré transformations, and dilatations, and conversely, which is true for no other ones. Moreover the set of these curves, straight lines and circles, and equilateral hyperbolas (and generalizations), is invariant under the conformal group, but not under larger ones, nor are there other curves that are so invariant. Thus the conformal group generates the points and fundamental curves of geometry, and leaves the set of them invariant.

Transversions then transform within, and leave fixed, the set of curves (surfaces in higher-dimensional spaces) that are created by the transformation group of geometry, the Poincaré group, and most important, keep the properties of (local) geometry the same (sec. III.1.d, p. 108).

And that there is no intrinsic length in geometry requires invariance under dilatations. Thus these conformal transformations describe the geometry (locally), and determine (completely?) the fundamental laws of the physics that takes place within it.

This holds for very general geometries, for it is only their local properties that are relevant. The restrictions hold for any manifolds, and it is probably an open question whether they hold more generally, and how general the geometry need, and can, be for them to hold. Since our space is at least a manifold (can it be otherwise?) the conformal group determines much (all?) of its geometry and physics.

I.2.c.ii *The geometrical significance of straight lines, circles and equilateral hyperbolas*

The group of conformal transformations selects a special set of curves, straight lines (including ones with zero length — points), circles or equilateral hyperbolas (including ones with infinite radii — straight lines), and likewise for other-dimensional spaces. Why geometrically are these special, why is geometry (given by its properties such as the sum of the angles of a triangle) invariant under transformations that leave these sets invariant, but only under such transformations? As we have just seen they define the points and coordinate systems of geometry. But there are additional reasons.

The path of an object moving on a circle (or equilateral hyperbola) is special, since a circle is special. All points on a circle (or sphere or hypersphere) are identical, which is not true for an ellipse, or other curves. And so it is for an equilateral hyperbola: there is no way of distinguishing one point on it from any other. The set of transformations that leave these curves and the sets of them invariant, such as rotations for Euclidean space, and generally the conformal group for 3+1

18

space, are therefore a fundamental property of the geometry — thus of physics; the conformal group, and no larger group, leaves the sets of these special curves invariant, as it does angles (sec. I.5.d, p. 45), and these are closely related.

Consider the difference between a circle (the Euclidean equivalent of an equilateral hyperbola) and an ellipse. For a circle, taking the origin at the center, we draw the x axis through any point. The transformation to the x' axis going to any other point is a rotation. The tangent at that point is given by the same rotation from the tangent at the first. For an ellipse, a rotation from x to x' does not give a point on the ellipse. Rotations for a circle and its inverse are equivalent, they go from one point on the circle to another on the circle — but not for an ellipse. A circle, with equation

$$\frac{x^2}{r^2} + \frac{y^2}{r^2} = 1,\qquad\qquad\text{(I.2.c.ii-1)}$$

has the slope given by the coordinate (angle) at any point, taking the x' axis from the center to that point and y' its perpendicular, the tangent, with

$$\frac{dy'}{dx'} = -\frac{x'}{y'} = -\frac{1}{tan\theta},\qquad\qquad\text{(I.2.c.ii-2)}$$

where θ is the angle between x and x'. For an ellipse with equation

$$\frac{x^2}{a^2} + \frac{y^2}{b^2} = 1,\qquad\qquad\text{(I.2.c.ii-3)}$$

the slope is given by

$$\frac{dy'}{dx'} = -\frac{b^2}{a^2}\frac{x'}{y'} \neq -\frac{1}{tan\theta}.\qquad\qquad\text{(I.2.c.ii-4)}$$

Thus the coordinate system at that point is not obtained by the rotation to the point.

For an ellipse a point is reached from another by a rotation and an elongation or contraction, a dilatation, but not of the whole plane, so the transformation is not a rotation. For a circle, if space is invariant under rotations, all points are equivalent, coordinate systems given by axes through the center, and those tangent to the circle and perpendicular to the tangent at any point, are equivalent, which is true of no other curves. For these, like ellipses, the rotation from the x axis to the x' axis, and that from the tangents to the curves at the points of intersection with these axes, are different. Rotational invariance of space does not place conditions on coordinate systems to different points on other curves, or tangents to different points, it does not require them to be equivalent.

Rationales are similar for equilateral hyperbolas (obtained using $y \Rightarrow it$): Lorentz transformations (rotations in indefinite metric spaces) and their inverses go from one point on an equilateral hyperbola to another. This is not true for other curves, other paths — for nonconstant accelerations. For hyperbolas that are not equilateral, coordinate systems at different points are related not by Lorentz transformations, but by these plus point-dependent dilatations.

For a (nonequilateral) hyperbola similarly obtained from an ellipse, a Lorentz transformation from the frame of an object with constant velocity (from a straight line) to a point on the hyperbola is not the same as that from the first frame to the tangent frame at that point (the one instantaneously at rest). Thus neither Lorentz transformation properly relates the frames — so physically they do not impose conditions. For an equilateral hyperbola (which is obtained from a circle), these transformations are the same, going from one point to an indistinguishable one, so with implications for physical laws. If geometry is invariant under Lorentz transformations, physical laws cannot distinguish what geometry makes indistinguishable. But invariance is not necessary for geometrical significance, and whether physics is invariant or not, it takes place in a geometry for which these curves are fundamental.

These curves (and in other-dimensional spaces, surfaces) are fundamental properties of geometry, determined by it and determining it. The conformal group is the group, and the largest group, whose transformations generate these surfaces, and leave the set of them invariant. Under this group, and under no larger one, the set of these defining surfaces goes into itself.

We thus see why the conformal group, so these curves, or perhaps conversely these curves, so the conformal group, bear preeminent relationships to geometry. But the group is significant for other reasons also.

I.2.c.iii   *Invariance of analytic functions*

The conformal group is the largest group of a space leaving invariant angles (it is conformal), and also, in a plane, the Cauchy-Riemann equations. These are the equations that (all) analytic functions obey. Thus a transformation not leaving these invariant takes an analytic function to one that is not analytic, so the conformal group in the plane is the largest leaving analytic functions analytic. And this can be generalized to other spaces (sec. II.5.f.ii, p. 104).

I.2.c.iv   *Invariance of commutation relations*

In addition to these reasons for the (physical) relevance of the conformal group, commutation relations between coordinates and momenta are invariant under all conformal transformations. This follows from invariance of angles. Momentum operators give infinitesimal displacements so have the realization

$$p_x = i\frac{d}{dx}. \tag{I.2.c.iv-1}$$

Thus for a set of curves forming a coordinate system, mutually orthogonal where they cross (for example straight lines), there are, from orthogonality, commutation relations of the form

$$[p_x, y] = 0. \tag{I.2.c.iv-2}$$

Under conformal transformations angles between tangents remain $\frac{\pi}{2}$, so

$$[p_x', y'] = 0, \tag{I.2.c.iv-3}$$

$$[p_x', x'] = 1. \tag{I.2.c.iv-4}$$

All such commutation relations are unchanged. But for other transformations angles are not invariant, so not all tangents are orthogonal, and commutation relations change.

I.2.c.v   The light cone and the conformal group

Also the group is the largest under which the speed of light, c, is constant [Gross (1964)]. Since in the proper units $c = 1$, what does this mean? Null equilateral hyperbolas are taken into null equilateral hyperbolas — the light cone is invariant under conformal transformations for finite points mapped into finite points [Wess (1960)]. This cone, in the z, t plane, forms two straight lines — angles between them and the axes are not changed. The conformal group, but no larger one, leaves angles invariant. For this angle,

$$tanh(\eta) = \frac{z}{t} = 1, \quad \text{so} \quad \frac{sinh^2(\eta)}{cosh^2(\eta)} = 1,$$

$$sinh^2(\eta) = cosh^2(\eta) = 1 - sinh^2(\eta), \quad sinh(\eta) = \frac{1}{\sqrt{2}}. \tag{I.2.c.v-1}$$

And the number of photons (photon norm) is invariant under the group.

I.2.c.vi   Conformal transformations between circles (or equilateral hyperbolas) and straight lines

There are conformal transformations, in the complex plane, that take circles and lines into each other (sec. II.3, p. 77) [Yaglom (1968)]. There is a dilatation taking any circle, or sphere, to one of any given radius; a line, or plane, is a circle, or sphere, of infinite radius. Thus invariance under the conformal group requires a certain equivalence of lines and circles.

Requirements imposed by one type of curve on physics means that there are requirements imposed by the other, so perhaps stronger and additional conditions. No larger group has equivalent properties — if lines (paths of constant velocity — zero acceleration) impose conditions these do not imply that there are conditions from paths of other shapes. Physical properties imposed by the necessity of being able to transform between observers with different constant velocity can produce ones from constant, but not nonconstant, accelerations.

The conformal group then is special, this set of rotations, boosts, translations, dilatations, and transformations to frames of constant acceleration, is fundamentally related to the geometry of space in which physics takes place. It thus can place restrictions on the laws governing that physics — these must be accord with the requirements of the underlying geometry.

I.2.c.vii *Physical implications of the existence of special curves of geometry; the spinning bucket*

There are in geometry special curves, circles, spheres, hyperspheres, in indefinite-metric spaces equilateral hyperbolas and hyperboloids, and the special cases of these, straight lines, planes and generalizations. They are fundamental properties of the geometry, and define it. What is their relevance for physics?

For an example consider the famous Newton's bucket experiment [Norton (1993), p. 799, 803]. If a bucket at rest holds water, its surface is flat, if it is spun, the surface curves. How does it know that it is being spun? One answer is a version of Mach's principle (which makes no sense) that inertia is determined by the presence of other matter in the Universe [Mirman (2001), sec. I.5.b, p. 22]. Thus in some magic way, this other matter informs the water that it is there, and causes its surface to curve. But the reason is really geometrical.

Straight lines and circles are properties of our geometry. And geometry does have properties — independent of physical objects, and such properties determine the nature of the physics that takes place within it, and they determine that physics can take place within it [Mirman (1995b)]. There are transformations that leave these curves invariant, and they are very different from ones that do not. For a triangle, the sum of its angles is π. Let us hope that no one suggests that the reason the angles add to π is because of the presence of other matter in the universe. If a triangle were of curved lines, the sum of its angles would be different — as it would if the space in which the triangle exists were curved (as on a sphere). This sum of angles comes from the geometry, only. And physical laws must distinguish between motions on the types of triangles, ones with different sums of angles — because the geometry does. Thus it also distinguishes, for the same reason, between motion along straight lines and that along equilateral hyperbolas (or circles), and also other motions — along other curves, that is with nonconstant acceleration.

The paths followed by each atom of water when the bucket is spun, and when it is not spun, are geometrically different, thus physically different, so the surface is different. That is why the surface curves. (Also of course there are interactions between molecules of water, and between these and the bucket and the external objects spinning the bucket.)

Physics must respect the geometry in which it takes place.

I.2.c.viii   *The path is determined by the interactions*

A constant acceleration, so along a path that is an equilateral hyperbola, is the result of all interactions, that is all observations made by the observer whose path it is. This implies that these are invariant under transformations giving the acceleration — all observations, hence physical laws, are so invariant. This is true for any constant acceleration (we can take the units so that any hyperbola is the unit one). This implies again that physical laws are invariant under transversions, so must be related by the conformal group, which places restrictions on them, although the real arguments, as we discuss, are stronger.

I.3   Transversions and constant acceleration

Perhaps the most interesting conformal transformations are transversions (at least at present, since they are relatively unfamiliar). They are different in many ways from other transformations that we may be more accustomed to, like those of finite and discrete groups [Mirman (1995a; 1999)], and rotation, Lorentz and Poincaré groups [Mirman (1995c)], so they help expand our understanding of geometry, group theory, the nature of realizations of group operators, the representations of groups, and (so) of physics. When appearing as transformations of the conformal group they are realized nonlinearly. Why nonlinearly, what is their geometrical significance, and what affect do they have physically?

The relevant question here is the last, with the others needed for full understanding of answers to it. Physically that transversions can be interpreted as transformations to frames with constant acceleration (sec. I.2.b.ii, p. 16; sec. I.3.b, p. 24) has long been known [Engstrom and Zorn (1936); Fulton, Rohrlich and Witten (1962a,b); Hill (1945,1947,1951); Page (1936a,b); Page and Adams (1936); Robertson (1936)], and has many ramifications. Why is constant acceleration, but only constant acceleration, special? The reasons are based on geometry, on certain curves of geometry being special — so paths of objects with constant acceleration are special. We must then show that these transversion transformations are to, and between, such frames.

Transversions are like boosts in changing one velocity to another, but for them velocity changes in time. If the initial acceleration is zero, transversions accelerate the object (correctly relate objects with different acceleration) — they go to an equilateral hyperbola from a straight line in z, t space (a special case of an equilateral hyperbola), or conversely (sec. II.3, p. 77; sec. II.3.d, p. 82).

Discussions here are purely kinematical, only considering paths and transformations of variables. Of course accelerated (charged) objects radiate and in a theory of them such aspects must be taken into account. This is not now relevant since we do not study causes of acceleration or how they are maintained constant. We just look at their paths and how the group relates these.

There is another interpretation of transversions: space-dependent rescalings, which we consider after the study of the acceleration interpretation (sec. I.5, p. 43).

I.3.a   We need consider only two-dimensional space

For constant acceleration there must be one point at which the velocity in the direction of the acceleration — taken as the z axis — is zero. For now we study motion only along a line — so confined to the z, t plane. Considering just two dimensions gives results that are largely general. (However because of dilatations the path with a velocity component that is not parallel to the acceleration is not simply composed of two independent motions [Jancewicz (1988), p. 270].) The path in this plane is an equilateral hyperbola.

Since we are interested in restrictions on laws of physics given by the conformal group (that is by geometry) we can consider only motion in a z, t plane, and we see from results so obtained that a constant velocity perpendicular to the plane does not change restrictions or results. We can choose the (inertial) observer so that it has the same velocity perpendicular to the z, t plane as the accelerated object, and thus that velocity can be taken zero, allowing our study to be restricted to planar motion.

I.3.b   Derivation of the equilateral hyperbola as the path of constant acceleration

To show that the path of an object with constant acceleration is an equilateral hyperbola (in a z, t plane), and conversely [Jancewicz (1988), p. 264], we take an object whose motion is one-dimensional (as is always possible for constant acceleration) [Anderson (1967), p. 200; Jancewicz (1988), p. 262; Rindler (1960), p. 39]. We consider then two derivations.

I.3.b.i *Acceleration and variation of momentum*

With u the (variable) speed as measured in a fixed frame (that of the observer who measures the acceleration), the momentum is

$$p = \frac{m_o u}{\sqrt{1 - u^2}},$$ (I.3.b.i-1)

where m_o is the rest mass as seen by an inertial observer (sec. I.1.c, p. 3). By constant acceleration, we mean that the rate of change of momentum with respect to time, measured by a fixed observer (time is the path parameter), divided by the rest mass, is constant. So, with m_o constant,

$$\frac{dp}{dt} = \frac{d(\frac{m_o u}{\sqrt{1-u^2}})}{dt} = m_o g,$$ (I.3.b.i-2)

$$d(\frac{u}{\sqrt{1 - u^2}}) = g\,dt,$$ (I.3.b.i-3)

with g a constant — the acceleration, which determines the hyperbola, is, of course, a constant along the path. Integrating, taking the origin of time at that point for which $u = 0$ (there must be one if the acceleration is constant),

$$\frac{u}{\sqrt{1 - u^2}} = gt,$$ (I.3.b.i-4)

so

$$u = \frac{gt}{\sqrt{1 + (gt)^2}} = \frac{dz}{dt},$$ (I.3.b.i-5)

where z is the position of the object as measured by the fixed observer, and t the time at which it has this position, measured by the observer's clock. Hence

$$z = \frac{1}{g}\sqrt{1 + (gt)^2},$$ (I.3.b.i-6)

and

$$z^2 - t^2 = \frac{1}{g^2},$$ (I.3.b.i-7)

an equilateral hyperbola. Here the sign of t^2 is negative since the metric is (1,-1).

From the origin the distance to point z, t on such a hyperbola is

$$d^2 = z^2 - t^2 = \frac{1}{g^2},$$ (I.3.b.i-8)

and is constant (as is the distance from the center of a circle to any point on it, and the origin can be on neither curve). So constant acceleration has the property that the distance between the origin and any point on its path is the same, no matter what the point. This is (only) somewhat similar to the light cone, for which the distance between any two points is zero.

If the acceleration is not constant the path is not a hyperbola. Other hyperbolas, besides equilateral, can be obtained by using unequal units for space and time (so that the speed of light is no longer 1, as is used here). This merely changes formulas, but with no physical difference.

An equilateral hyperbola is comparable to a circle in that it has all points identical. It is obtained from the circle

$$z^2 + y^2 = r^2, \tag{I.3.b.i-9}$$

by substituting $y \Rightarrow it$. If the initial velocity is not parallel to the acceleration the path is like a (hyperbolic) helix, but not quite the same (sec. I.3.a, p. 24).

We can write

$$t = \frac{1}{g}sinh(g\tau), \quad z = \frac{1}{g}cosh(g\tau), \tag{I.3.b.i-10}$$

$$\tau = \frac{1}{g}arctanh(\frac{t}{z}), \tag{I.3.b.i-11}$$

where τ is the path parameter — as the path is one-dimensional only one parameter is necessary to give a point on it.

I.3.b.ii *What is the path if the acceleration is zero?*

The equation for the path contains $\frac{1}{g^2}$, the inverse of the acceleration. But this should hold for all accelerations, certainly zero. In deriving this we took the initial time at the point at which velocity $u = 0$. This is not useful with zero acceleration. However if the velocity is zero at some time, we can always choose that as the initial time. But putting $g = 0$ gives

$$\frac{d(\frac{u}{\sqrt{1-u^2}})}{dt} = g = 0, \tag{I.3.b.ii-1}$$

$$\frac{u}{\sqrt{1 - u^2}} = c_u, \tag{I.3.b.ii-2}$$

so

$$u = \frac{c_u}{\sqrt{1 + c_u^2}}, \tag{I.3.b.ii-3}$$

a constant, as it should be. As

$$c_u \Rightarrow \infty, \quad u \Rightarrow 1, \tag{I.3.b.ii-4}$$

(the speed of light), and as

$$c_u \Rightarrow 0, \quad u \Rightarrow 0. \tag{I.3.b.ii-5}$$

Thus c_u gives the speed u in terms of that of the speed of light — which is really (geometrically) the slope of the line for which

$$z^2 - t^2 = 0 \qquad \text{(I.3.b.ii-6)}$$

— and this is (always chosen as) 1.

What is the path parameter, and how are z and t related to it? Since

$$z = vt, \qquad \text{(I.3.b.ii-7)}$$

we can take either z or t as the required parameter.

I.3.b.iii *Relating constant acceleration to change of velocity*

There is another derivation of an equilateral hyperbola. For an object with constant 4-acceleration a_μ we have [Misner, Thorne and Wheeler (1973), p. 166]

$$a_\mu = \frac{du_\mu}{d\tau}; \qquad \text{(I.3.b.iii-1)}$$

u is the velocity, τ the proper time (path length, or path parameter), and

$$a_\mu a^\mu = g^2, \qquad \text{(I.3.b.iii-2)}$$

which is taken as constant. Since

$$\sum p_\mu^2 = m^2 \sum u_\mu^2 = m^2, \qquad \text{(I.3.b.iii-3)}$$

$$\sum u_\mu^2 = 1; \qquad \text{(I.3.b.iii-4)}$$

the magnitude of the four-velocity is a constant, so the four-velocity and four-acceleration are orthogonal [Jancewicz (1988), p. 262],

$$a_\mu u^\mu = 0. \qquad \text{(I.3.b.iii-5)}$$

The motion is planar. Differentiating we get,

$$\frac{dt}{d\tau} = u^o, \quad \frac{dz}{d\tau} = u^1, \quad \frac{du^o}{d\tau} = a^o, \quad \frac{du^1}{d\tau} = a^1. \qquad \text{(I.3.b.iii-6)}$$

Combining equations (with the speed of light = 1 so z, t, τ and $\frac{1}{g}$ have the same units), and solving we obtain the equations of the path,

$$t = \frac{1}{g} sinh(g\tau), \quad z = \frac{1}{g} cosh(g\tau), \qquad \text{(I.3.b.iii-7)}$$

and

$$z^2 - t^2 = g^{-2}. \qquad \text{(I.3.b.iii-8)}$$

The path of an object with constant acceleration is thus an equilateral hyperbola, in agreement with the previous derivation.

For $g\tau$ small,

$$gz = 1 + \frac{1}{2}g\tau^2, \quad t = \tau, \quad gz = 1 + \frac{1}{2}gt^2, \qquad \text{(I.3.b.iii-9)}$$

as we expect.

I.3.b.iv *Acceleration of a particle moving in a circle*

It is more familiar to consider a particle moving in a circle for which we say that it has constant acceleration if the rate of change of velocity along the arc, with respect to arc length, is constant. This is equivalent to what we have obtained with hyperbolas. The linear velocity along a circle, with radius r, is

$$v_l = \frac{dl}{dt} = r\frac{d\theta}{dt} = r\omega. \tag{I.3.b.iv-1}$$

Constant (here angular) acceleration means that $\frac{dv_l}{d\theta}$ is constant — the acceleration is constant along the circle — with position given by θ, the path parameter which runs from $-\infty$ to ∞. So

$$\frac{dv_l}{dt} = \frac{dv_l}{d\theta}\frac{d\theta}{dt}. \tag{I.3.b.iv-2}$$

$$\frac{dv_l}{d\theta} = r\frac{d\omega}{\omega d\theta} = \text{constant} = \alpha. \tag{I.3.b.iv-3}$$

Coordinates x, y are then

$$x = r\cos\theta = r\cos(\frac{l}{r}), \tag{I.3.b.iv-4}$$

$$y = r\sin\theta = r\sin(\frac{l}{r}), \tag{I.3.b.iv-5}$$

with l the path length. This is equivalent to the expression for the hyperbola: radius r corresponds to $\frac{1}{g}$, the inverse of the acceleration, and path length l to τ. When the acceleration goes to zero, $\frac{1}{g} \Rightarrow \infty$, the hyperbola becomes a straight line, as the circle does for $r \Rightarrow \infty$, indicating why they correspond.

I.3.c **Energy and momentum along the path**

These derivatives give the momentum, seen in an inertial frame,

$$p = m_o g t; \tag{I.3.c-1}$$

it depends linearly on time, as expected. The energy, with one-dimensional motion, is defined by

$$E = \sqrt{m_o^2 + p^2} = m_o\sqrt{1 + (gt)^2}, \tag{I.3.c-2}$$

and

$$\frac{dE}{dt} = \frac{m_o g^2 t}{\sqrt{1 + (gt)^2}} = m_o g u. \tag{I.3.c-3}$$

Integrating from $t = 0$, when $E = m_o$ (so $p = 0$), to t,

$$E = m_o g^2 \int dt \frac{t}{\sqrt{1 + (gt)^2}} = \frac{1}{2} m_o \int d(gt)^2 \frac{1}{\sqrt{1 + (gt)^2}}$$

$$= \frac{1}{2} m_o \int d\zeta \frac{1}{\sqrt{1 + \zeta}} = m_o \sqrt{1 + \zeta} - C = m_o \sqrt{1 + (gt)^2}. \quad \text{(I.3.c-4)}$$

Hence

$$E^2 - p^2 = m_o^2 = m_o^2 (1 + (gt)^2 - (gt)^2), \quad \text{(I.3.c-5)}$$

which remains constant (independent of position on the path), as it must being a Poincaré invariant. And it is independent of acceleration.

As we know (eq. I.1.c-2, p. 4), and as we see again, this equation for the invariant, including the mass value, is a property of the object. Interactions do not affect the mass (invariant), but rather the expressions for momenta.

I.3.c.i *Energy and momentum using the path parameter*

We can write the expressions for energy and momentum in terms of path parameter τ (eq. I.3.b.i-10, p. 26; eq. I.3.b.iii-7, p. 27),

$$p = m_o gt = m_o sinh(g\tau), \quad \text{(I.3.c.i-1)}$$

$$E = m_o \sqrt{1 + (gt)^2} = m_o \sqrt{1 + (sinh(g\tau))^2}$$
$$= m_o cosh(g\tau),$$

$$\text{(I.3.c.i-2)}$$

$$E^2 - p^2 = m_o^2 = m_o^2 (1 + sinh(g\tau)^2 - sinh(g\tau)^2), \quad \text{(I.3.c.i-3)}$$

For an object starting at rest, the energy at $t = 0$ is the rest mass, and energy and momentum increase indefinitely as we expect for an accelerated object, but the invariant always equals m_o^2.

Also like the path, the curve in the E, p plane (the dual space) is an equilateral hyperbola.

An equilateral hyperbola, through a given point, is determined by a single parameter, this interpreted as the (inverse) acceleration, or mass (in the dual space). This defines the particular hyperbola, and is the same (measured) at all points.

The rest mass of an object with constant acceleration observed by an inertial observer does not depend on acceleration. It, seen by an observer moving with constant velocity, is independent of the velocities of observer and object and the object's constant acceleration. A Poincaré invariant, it is the same for all straight line paths, and also for equilateral hyperbolas. This indicates again that these two types of curves are closely related and can be considered as a set.

I.3.c.ii *The mass of an object with uniform acceleration simulated by gravity*

What is the mass, m_a, measured by an observer with constant velocity which at rest has value m_o, of an object with constant acceleration [Jancewicz (1988), p. 268]? We can take the uniform acceleration to be simulated (approximately) by a constant gravitational field [Fulton, Rohrlich and Witten (1962a), p. 456], giving

$$m_a = m_o(1 + gh), \qquad\qquad (\text{I.3.c.ii–1})$$

where g is the gravitational field, and h is the height. Take the observer to be instantaneously at rest with respect to the object, and coinciding with it, and also zero potential to be at that point. Thus we find again

$$m_a = m_o. \qquad\qquad (\text{I.3.c.ii–2})$$

As we know the mass is constant in time (so our choices are general), and further the proportionality factor (here $1 + gh$) is independent of the mass. Mass ratios are invariant. But this simulation is really good only at one point.

I.3.d   Accelerated statefunctions

Conformal transformations are defined over space, and act on its co-ordinate systems, with subgroups that rotate or translate coordinates. These (with generators in parentheses) acting on statefunctions, translate them (the P's), rotate the direction of the three-momentum (M_{ij}), change its magnitude (M_{4i}), and scale (D), and for the K's (the Lie-algebra transversion operators), take frames with uniform acceleration to other such ones, so relate ones with different constant acceleration.

What is the form of an accelerated statefunction, what general properties does it have? Choice of basis states is, within wide limits, arbitrary. For objects with constant velocity we pick states of the form $exp(ipx)$, where the x's are the (3+1) Cartesian coordinates, and p is a (schematic symbol for a) constant (vector), or sums of these functions, wavepackets. We want reasonable basis states for accelerated objects. These states are not those found by transversion transformations on states for constant velocity. That is like taking an object with constant velocity and accelerating it; here we study objects with constant acceleration.

I.3.d.i *Statefunctions for constant velocity*

Our choice of statefunctions of accelerated objects is guided by the choice for objects in inertial frames, so we mention these. With constant velocity these are (schematically),

$$|v_{constant}) = exp(ip_z z + iEt) = exp\{it(p_z v + E)\}, \qquad \text{(I.3.d.i-1)}$$

$$E^2 - p_z^2 = m^2. \qquad \text{(I.3.d.i-2)}$$

By choosing the origin differently we change this by a phase, which thus has no physical significance.

I.3.d.ii *How conformal transformations act on group basis vectors*

To study what the accelerated statefunctions should be, we summarize the effect of the conformal group on these, its basis vectors. We consider state

$$|s) = N exp(i\underline{p} \bullet \underline{x}), \qquad \text{(I.3.d.ii-1)}$$

where N is the normalization. A rotation of both p and x, that is of the object and the observer, leaves this unchanged. However we can rotated either one to get

$$|s)' = N exp(i\underline{p} \bullet \underline{x}'). \qquad \text{(I.3.d.ii-2)}$$

It does not matter which is rotated, the final state is the same, but differs from the original. However if space is rotationally invariant, the transformed state is physically equivalent to the untransformed one — they cannot be distinguished experimentally, only the effect of their orientations with respect to an observer has physical consequences. For a dilatation the analysis is similar. Transforming both p and x leaves the state the same. A transformation of p gives the state

$$|s)' = N exp(i\lambda \underline{p} \bullet \underline{x}), \qquad \text{(I.3.d.ii-3)}$$

while for x,

$$|s)' = N' exp(i\underline{p} \bullet \underline{x}/\lambda), \qquad \text{(I.3.d.ii-4)}$$

with the normalization changed since probability is proportional to $N^2 dx$, and dx is a physical object, say a very small rod. If the length of that is changed, the normalization must be since we want the probability of finding a particle on the rod to be unaffected.

Is the state obtained by $p \Rightarrow \lambda p$ physically different? Since there is no length in the geometry, there is no way to distinguish the states. All that can be found are relationships to other objects, or to a set of them like the objects in the universe. If we considered the state in isolation, we might say that multiplying by λ makes the momentum larger. But larger compared to what?

In these cases both the object and observer can be transformed, with no discernible difference. Or one can be transformed, but it does not matter which. It is only the relationship between them that is relevant. Transversions are different. We can consider an object both inertial and accelerated, and see how the transformation affects its properties. However an accelerated observer is different from an inertial one (sec. I.1.d.iv, p. 7), so unlike the other transformations we do not have a choice of whether to transform the object or the observer, obtaining for them physically indistinguishable situations. Here the two situations are physically different.

Transversions are nonlinear transformations. For the other group elements we can transform x and perform the reverse transformation on p to keep $\underline{p} \bullet \underline{x}$ constant. But if we tried to do that for transversions we have two choices. The coefficients in the transformations of p could be functions of x, but then they would not be transversions, so not members of the conformal group. Otherwise we could take the transformations of x and p identical, with the coefficients in the former functions of x, for the latter of p, giving transversions on x and on p. But then $\underline{p} \bullet \underline{x}$ would not be invariant.

Also we can leave p unchanged and replace x by the function for an accelerated object, an equilateral hyperbola. This however is not what we want: functions that are constant over the path, except for a phase. Such functions are $exp(ipxcos\theta)$ and $exp(ipxcosh\theta)$. These are for objects moving on straight lines. If we replace the line by a hyperbola (or a more easily visualized circle) p is constant as the object moves, but the cosine of the angle between p and $x, cos\theta$ (or $cosh\theta$), varies. Hence the change in the function is not merely in its phase.

Choice of basis states is arbitrary, picked to satisfy certain conditions, in these discussions to give representations that are also Poincaré representations with momentum diagonal. So to decide what to pick for accelerated motion we consider what momentum-diagonal statefunctions are with constant acceleration, and why they have the form that we specify.

I.3.d.iii *Why uniformly accelerated statefunctions are momentum eigenfunctions*

Representations considered here are limited to those whose bases are momentum eigenfunctions, belonging to the best-known Poincaré representations [Mirman (1995c), chap. 2, p. 12]. However we also want them to be basis vectors of the entire conformal group (for the type of representations we are studying). Why are the proper accelerated statefunctions momentum eigenfunctions or sums of these, wavepackets? And what is meant by a momentum eigenfunction if the path in z, t space is curved?

Momentum eigenfunctions and only these are invariant under translations — because the momentum operator is (exponentiated) the translation operator. Thus for an object traveling with constant velocity all points on its path are identical: $exp(i\underline{p} \bullet \underline{z})$ is invariant under changes in z, except for an irrelevant phase. Physically this means that the probability of finding the object is the same at all points, which is not true of a wavepacket, although this can be expanded in terms of momentum eigenfunctions.

An object with constant acceleration moves on an equilateral hyperbola, and this is also identical at all points — motion along it, changing the path parameter, does not produce relevant change (phase is irrelevant). The probability of finding the object is the same everywhere on the curve.

Consider the difference between a circle and an ellipse. A conformal transformation, in a space of positive-definite metric, takes a line — all points are identical — to a circle (chap. II, p. 66), also with no way of distinguishing points. For an ellipse, however, points are distinguished. An equilateral hyperbola (for an indefinite-metric space) is equivalent to a circle.

Thus a basis vector must be such as to not distinguish points on such hyperbolas. But the only such functions are of the form $exp(ip\tau)$, where τ is the path parameter — momentum eigenfunctions. This momentum however is not that along coordinate axes, or sums of such. The momentum generator is not $i\frac{d}{dx}$, but $i\frac{d}{d\tau}$. And τ is not a sum of coordinates, rather they are hyperbolic functions of it (eq. I.3.b.i-10, p. 26).

Therefore the proper statefunctions on paths of constant acceleration are momentum eigenfunctions. However this momentum operator is a sum of momentum operators in the z and the t directions, giving a translation at an angle to these, but with this angle changing along the path. Thus these statefunctions are momentum eigenfunctions for the translation operator along the path, but not for translation operators along z or t.

Statefunctions of uniformly accelerated objects, seen by observers at rest, are then, in elliptic coordinates (sec. II.4, p. 96),

$$|\text{accelerated}\rangle = exp(ip\tau + iq\sigma + ip_x x + ip_y y), \qquad (\text{I.3.d.iii-1})$$

where σ is the coordinate orthogonal to the path, and τ is, as we know, a nonlinear (hyperbolic) function of z and t; x and y are the rectangular coordinates perpendicular to the z, t plane. Since these objects move along a hyperbola, $q = 0$ — the statefunction does not depend on q. And the values of the momenta are eigenvalues of an operator containing nonlinear terms.

For nonconstant acceleration, we would not expect that the probability to be the same at all points of the path. Variable acceleration changes the likelihood of an object being on parts of the path that are reached with increasing time.

Hence the proper choice for the functions on which to build the conformal group representations is this set of momentum eigenfunctions, exponentials in the path parameter, or sums of such.

Here then is an outline of how conformal group representations should be constructed. We shall not work out details, but merely note that they have a larger set of basis states than those only for constant velocity. Thus the representations are likely to be more complicated, but perhaps useful in different ways, than are the more restricted ones.

I.3.d.iv *Statefunctions with acceleration*

To choose statefunctions (basis states) for constant acceleration, and see if they describe the required behavior of these objects, we take velocity and momentum to be zero at $\tau = 0$, so using (sec. I.3.c.i, p. 29)

$$p = m_o sinh(g\tau),\qquad\text{(I.3.d.iv-1)}$$

$$E = m_o cosh(g\tau),\qquad\text{(I.3.d.iv-2)}$$

$$t = \frac{1}{g}sinh(g\tau),\quad z = \frac{1}{g}cosh(g\tau),\qquad\text{(I.3.d.iv-3)}$$

$$\tau = \frac{1}{g}arctanh(\frac{t}{z}),\qquad\text{(I.3.d.iv-4)}$$

$$t_o = 0,\quad z_o = \frac{1}{g},\quad E_o = m_o,\quad p_{oz} = 0.\qquad\text{(I.3.d.iv-5)}$$

The statefunction is then

$$|p) = Nexp(ip\tau) = Nexp\{\frac{ip}{g}arctanh(\frac{t}{z})\}.\qquad\text{(I.3.d.iv-6)}$$

For a momentum eigenfunction motion along a path should change only the phase. This clearly is true for these functions.

On lines perpendicular to paths basis functions are constant, (equivalent to) plane waves. Statefunctions are defined over the entire z,t plane, but are functions only of τ. The momentum operators are $i\frac{d}{d\tau}$, with eigenvalue p, and $i\frac{d}{d\sigma}$, where σ is perpendicular to τ, with eigenvalue 0. Generalizations to three (or more) dimensions are immediate.

The argument $\frac{t}{z}$ varies from 0, at the time origin, to 1, at the light cone. Taking derivatives with respect to the rectangular coordinates we see that as it approaches the light cone the basis function oscillates more and more rapidly. Why? For an accelerated object the momentum, energy and frequency increase, the wavelength decreases. And so, as we expect, the rate of oscillation, as a function of z or t, increases. This basis vector does properly describe accelerated motion.

Transversions take one equilateral hyperbola to another thus leave the form of these basis vectors invariant. However the variable (state label) $\frac{p}{g}$ is changed, because the acceleration is (sec. II.3.e, p. 88). The specific basis state is frame-dependent as it is for plane waves that are basis states for constant velocity. This frame dependence is given by p.

These basis states are analogs of plane waves (and as functions of τ they are plane waves). However there are other types of representations of inhomogeneous groups using as basis states spherical harmonics, or generalizations of them to other spaces. For a space with an indefinite metric these are obtained by replacing the proper sines and cosines by their hyperbolic equivalents. As we know for the Poincaré group these functions can be expanded in terms of each other (sec. A.6, p. 240). The choice of representation type depends on the application. For the conformal group the representation functions that we have depend on τ, so after we find the equivalent of spherical harmonics for this variable, we can substitute its expression in terms of rectangular coordinates. This would then give the analogous basis states, which would be different expressions then the generalizations of spherical harmonics to indefinite metric spaces. Their properties have to be investigated. In particular sets of spherical harmonics are intermixed by the rotation group, but not mixed with other sets, each set being a different representation. It has to be determined whether these generalizations form analogous sets, that is whether there are different representations for the conformal group.

I.3.e Why constant acceleration is related to the conformal group action on space

Why geometrically are transversions and constant acceleration related? Special conformal transformations are products of translations with the interchange of the inside and outside of an (n-dimensional) spherical surface, or for a space with indefinite signature, an equilateral hyperboloidal surface (the equivalent of a hypersphere), leaving the surface itself (the subspace) invariant (sec. III.1.d, p. 108). An object with uniform acceleration moves on such a hyperboloidal surface. Thus its trajectory is invariant under a transversion. For each such trajectory there is a transversion that leaves it invariant, and conversely. Moreover the set of spheres with centers at the origin, or hyperboloids, are all parallel so if a transversion leaves one fixed, then the others remain parallel to it, that is spheres or hyperboloids. These are the paths for constant acceleration, so the set of these paths is invariant.

This is the geometrical property of transversions that causes them to be related to constant acceleration, and the geometrical property of paths of constant acceleration that relates them to transversions.

Motion with constant momentum is taken into motion with constant momentum by the Poincaré subgroup, and paths of uniformly accelerated motion are taken into paths of uniformly accelerated motion by special conformal transformations. Thus the set of paths of constant acceleration (including zero) is invariant under the group — the set of equilateral hyperbolas in an indefinite metric space, including straight lines (those parallel to the t axis are for objects at rest) are so invariant. These are the most general curves that are.

Properties of geometry are thus related to the conformal group. However there are (apparently) no such properties of space, or any (nontrivial) subspaces, invariant under larger groups.

I.4 The conformal group and the Poincaré representation

Fundamental physical objects are (specified by) basis states of irreducible Poincaré representations [Mirman (2001), sec. I.8.b, p. 48], labeled by the two group representation invariants, rest mass and spin at rest (for massive objects) — specifying the rest-frame spin and mass is sufficient to determine that the object is, for example, a free electron. What happens to these under the conformal group, what effect does the group have on Poincaré representations? Dilatations leave mass ratios and spin unchanged, what about transversions?

Inertial observers at rest with respect to an accelerated object and at the same point — the origin of space and time, see the same rest mass (sec. I.3.c.i, p. 29) as for zero acceleration [Fulton, Rohrlich and Witten (1962b), p. 665]. This is not completely unexpected as there is no way of distinguishing objects with and without acceleration if they are instantaneously at rest and at the same point as an observer — however a measurement cannot be done in zero time, so variation of acceleration is relevant. An electron is an electron accelerated or not. However as the object moves away its properties and behavior, such as speed, energy and momentum, measured by an observer at rest, depend on its (constant) acceleration. But energy and momentum obey a constraint so the rest mass, given by $E^2 - p^2$, is constant.

This we have to now consider.

I.4.a The spin label

One Poincaré invariant determines the spin of an object (an internal variable). Representations of the rotation subgroup (actually its SU(2) covering group) are labeled by angular momentum, and since the type of object is independent of its motion (an electron is an electron no matter what its velocity or acceleration), as are values of its internal variables, that type is identified by its angular momentum at rest, its spin. The (covering group of the) Lorentz subgroup representation is labeled by its smallest angular momentum value, the spin in the rest system [Gel'fand, Minlos and Shapiro (1963), p. 188; Naimark (1964), p. 116]. The other Lorentz label (sec. A.2, p. 224) is irrelevant here since we are considering the Poincaré subgroup of the conformal group [Mirman (1995b), sec. 6.2.2, p. 110].

The magnitude of spin (in the rest frame), the Lorentz subgroup representation label, is clearly invariant under dilatations; angular momentum and dilatation operators commute (sec. I.1.e, p. 8). So we need look only at transversions.

Representations of the rotation subgroup are discrete; their label has discrete values only. Spin then has only discrete values, as is quite well-known. Acceleration however is continuous, it can have any value (and transversions, no matter how interpreted, depend on continuous variables). Thus if this representation label depended on acceleration it would have to also be continuous, which it is not. Since the spin representation label is discrete, and as the acceleration becomes zero it must go smoothly to the proper value for constant velocity, it has to be invariant — the same as at rest.

There is another argument for invariance of spin under transversions. These can be written as products of translations and an inversion (sec. III.1.d, p. 108). Translations do not change spin, and the inversion cannot take the spin value into a different one (the inversion cannot change spin-$\frac{1}{2}$ to spin-$\frac{3}{2}$, for example). The space obtained by a transversion transformation is also invariant under the Poincaré group, thus under its rotation subgroup, so conditions on spin values are not changed. But transformations of a group larger than the conformal group do not leave space Poincaré-invariant.

Also, since spin is an internal variable it must remain constant along the path.

For uniform acceleration then spin must have the same value as for zero acceleration — an object's spin seen by an inertial observer instantaneously at rest with respect to it is independent of the acceleration of the object. An observer not at rest would have to determine the complete statefunction, expand it into angular momentum eigenstates, and find that with smallest angular momentum. The argument is the same for an observer with constant acceleration which is also continuous, while spin is discrete. So an electron, say, has the same spin (but not angular momentum) as it does at rest, if accelerated, and if measured by an observer with a constant acceleration (who would however have to make different, and more complicated, measurements).

However while spin is invariant, total angular momentum is not. This is a sum of angular momentum states with coefficients that depend (continuously) on acceleration and velocity, and go to their proper values as these become zero. Experimentally the spin has to be extracted from the measurement of the angular momentum of an accelerated object.

What meaning does spin have then? Experimental determination of the expansion of the total angular momentum in terms of irreducible SU(2) representations (using procedures that we do not study here) gives these coefficients, and the smallest value of the angular momentum in such a sum is the spin.

I.4.b   The mass of an object seen by an accelerated observer

The second Poincaré label is physically interpreted as rest mass. And mass is invariant along the path (sec. I.3.b, p. 24) for equilateral hyperbolas. Transversions are here interpreted as transformations from a frame with one acceleration (including zero) to that of another (possibly zero). What affect do these have on mass?

This question is ambiguous so we must specify it, actually them. We start with rest mass, that measured by an inertial observer transformed back to the system in which the object is at rest. What is measured by that observer instantaneously at rest with respect to an object with uniform acceleration?

Another series of questions asks how answers to these vary along the path, how do these various observers, at rest with respect to the object at different points differ, if they do, in the rest masses that they determine? Since laws of physics can (presumably) be fully determined by inertial observers, we consider only these, and objects that have constant acceleration.

The answers to these we know: the mass is constant along the path, and independent of acceleration, so remains the rest mass (sec. I.3.c, p. 28).

We might also ask how mass changes if an object with constant velocity, say at rest, is accelerated? However this is not relevant here. Since the acceleration is always taken as constant, the object can never have constant velocity, cannot be at rest for a finite time (one reason that constant acceleration differs from nonconstant acceleration). We are not studying acceleration, but rather comparing objects with different accelerations (including zero).

I.4.c   The conformal group leaves the Poincaré representation unchanged

An observer (no matter what its position) at rest with respect to an object at some time finds the Poincaré representation of which the statefunction is a basis vector the same as if they were both inertial, but the particular basis vector itself is different. And this representation does not vary along the path; the representation it finds at any later time is the same.

As we see the mass is the same for all constant accelerations, including zero. The definition and properties of an elementary object (one given by an irreducible representation of the Poincaré group, labeled by mass and internal variables — for the Poincaré group only spin) are not changed by constant acceleration; the Poincaré representation is independent of it.

Under conformal transformations, and no others, functions that are rotationally invariant (or Lorentz invariant for a hyperbolic geometry) remain so. Thus these preserve fundamental invariances of space (sec. I.2.b, p. 15), so laws of physics that are Lorentz invariant must be invariant under these transformations. Also translational invariance goes into rotational (or Lorentz) invariance, and conversely (sec. II.3, p. 77).

Poincaré invariant equations, the laws of physics, go to Poincaré invariant equations. And this is true for no larger group.

There are several intuitive arguments for invariance of the Poincaré representation.

I.4.c.i *Time translation and the effect of transversions*

The mass of an object is invariant (for constant acceleration), the same measured in every inertial system instantaneously comoving with the object at some time. These differ only in being comoving at different times — being at rest with respect to the object at different times, at different points along the path. Also spin can be measured by these observers since they are instantaneously at rest with respect to it.

Physics however is time-translationally invariant. Thus all these frames must be equivalent, so all must measure the same rest mass — it is invariant under transformations from frames tangent to the path of constant acceleration to any other such frames, the comoving frames instantaneously at rest with respect to the accelerated object at any time. There is no way of picking a special time, or frame.

This is not true for nonconstant acceleration — the shape of the path changes in time. Thus we can pick a time that is distinguished, say that at which the acceleration is zero. So frames, which are labeled by (a meaningful) time, need not be, and are not, equivalent.

I.4.c.ii *Lorentz transformations along equilateral hyperbolas*

Inertial coordinate systems instantaneously at rest with respect to a uniformly accelerated object, whose paths are tangent to the path of the object, but at different points, thus with different constant velocities (as seen in a fixed system), are related by Lorentz transformations. Physics is invariant under Lorentz transformations so all these systems are equivalent. This, as does invariance under time translation, indicates why physical laws for an object with constant acceleration are the same as those for one with constant velocity.

For other curves for which such coordinate systems are not related by Lorentz transformations, but by, say, these plus point-dependent dilatations, this is not true. And physics is not, need not be, invariant under the latter — although transversions can be so interpreted (sec. I.5, p. 43), allowing such invariance, but these restricted transformations are not observable. Laws would then depend on the particular point-dependent dilatation, parameters would be functions of them, so laws for other curves are system-dependent.

As we see rest mass, measured by an inertial observer, who is at one time coincident with, and at rest with respect to, an object is independent of the (constant) acceleration for all time (sec. I.3.b, p. 24; sec. I.4.b, p. 38). Since all inertial observers are equivalent, this must be true for all. As the object has constant acceleration (so the motion can be taken to be in only one space dimension), it is at rest with respect to every inertial observer (moving in that direction) at some time. And properties of an object are independent of the position of the observer. Thus an observer finds, when at rest with respect to the object, that its mass is the same as if both had no acceleration. But the mass seen is independent of position along the path, so the observer sees the rest mass as constant — if it measures the mass at any time the value it obtains is that given by the fixed rest mass and the velocity it sees at that time (determined by the Lorentz transformation from the frame of the observer to that of the mass). It is independent of the constant acceleration (but, of course, the time, or path position, at which it has that velocity depends on its acceleration).

Transformations between systems, one with constant velocity, the other instantaneously at rest with respect to an object with constant acceleration (or one moving with constant velocity instantaneously coinciding with this), one whose path is tangent to the hyperbola, are Lorentz transformations, as are those between the second frame and the first. Physics is invariant under Lorentz transformations, thus all points on the path look the same [Jancewicz (1988), p. 265]; all reference frames instantaneously at rest with respect to the object but at different points of the path are equivalent.

Physics is invariant under Lorentz transformations as our (pseudo-) Euclidean geometry is invariant under these rotations, so giving consequences, as we see, for conformal transformations — emphasizing again the relationship of these transformations to geometry.

Why are time, translational and Lorentz invariance related, as this implies? If the properties of an object with constant acceleration, and the laws governing them, were not invariant under all three, they would be invariant under none. These would change with speed, but this is position and time dependent, so change under one means a change under all.

I.4.c.iii *Transversions imply interactions*

While mass as measured in an inertial frame may seem fundamental, an object with constant acceleration is never in such a frame, and it is an additional assumption that the object is identical to that in an inertial frame (sec. I.1.c, p. 3). Thus for an electron with constant acceleration we assume that it is in some basic sense the same type of object as an electron at rest. One argument for this is that it can be given a nonconstant acceleration, which is then set to a constant, so the electron can be followed, and other particles with that same constant acceleration can be compared with it, identifying them as the same type object.

In addition a constant acceleration can have any value (including zero) so if we considered the type of object to depend on the acceleration, there would be a continuously-infinite number of types of objects, a continuously-infinite number of types of electrons, say.

Accelerations mean that there are interactions. Thus conformal transformations require nonlinearities; these appear, but in a different way, in the realization of the K's (sec. III.3.a, p. 120). Conformal transformations therefore imply interactions (except, of course, if the parameters for the transformations by the transversions are zero) – these relate accelerated frames, and acceleration requires interactions.

The Klein-Gordon equation (an expression of the Poincaré invariant giving the mass [Mirman (1995b), sec. 6.3.2, p. 116]) and Dirac's equation (giving the mass and spin, but for spin-$\frac{1}{2}$ only) are thus invariant. For interactions, there is little freedom, these being strongly constrained (perhaps completely constrained) by the Poincaré group and dilatations — the similitude group. Also they must reduce to the ones seen by an observer with constant velocity if the acceleration is zero. Thus their fundamental forms also are invariant under the conformal group, as is true for the two Poincaré invariants with interactions inserted, for any spin. Laws of physics, given by these, are thus invariant under the conformal group. And transversions can be written as products of translations and an inversion (sec. III.1.d, p. 108), so interaction terms must be unchanged by them since they are by these transformations.

While the Poincaré representation of an object is not changed by the conformal transformations, the transversions imply changes in the expressions for the Poincaré invariants (adding terms, interactions), with these, acting on transformed basis vectors, retaining their values.

I.4.d   Consequences of constant acceleration come from the full theory

Questions have been raised about the meaning of constant acceleration, such as does a charged particle with a constant acceleration radiate? However discussions of these questions are classical, thus merely
interesting.

We know the equations governing charged particles, the Poincaré invariants [Mirman (1995b), sec. 6.2, p. 110] (Dirac's equation for spin-$\frac{1}{2}$),
so the proper way of considering such questions is by solving the equations, with constant acceleration, or constant gravitational field. There
is no reason to believe that there will be disagreements, or ambiguity,
if this is done correctly.

Likewise the equivalence principle can lead to interesting guesses,
but since it is a consequence of general relativity, application of all
equations of the theory, which are known, must give correct results,
unless these are inconsistent which is unlikely. Equivalence principles
are corollaries of the full theory. They should not be elevated to su-
perstitions, as, for example, the uncertainty principle often is [Mirman
(2001), sec. V.7.f.iii, p. 261]. And it is not necessary, and is incorrect, to
guess their meaning. That follows from the theory.

When all equations of a theory are known, placing additional requirements (except when unavoidable for calculational purposes) is not
only unnecessary but can lead to inconsistencies. General relativity and
quantum mechanics need no such beliefs or presumptions, nor do they
allow any.

The equivalence principle, for example, was a fundamental heuristic
tool in developing general relativity, but is not a fundamental assumption of it. Once a theory is known, the heuristics used for finding it
can either be derived from it, or are irrelevant or wrong. Yet here, as
so often happens, physicists believe that because a concept led to the
development of a theory, the theory is based on that concept.

For the conformal group we can obtain equivalent consequences (but
still have not). It is the group, and the largest group, under which laws
of physics are invariant. Two objects, one moving with constant velocity, the other with constant acceleration, and there are such possible
ones, for any velocity and any acceleration, obey the same fundamental laws. Hence laws are restricted by the requirement that they be so
invariant. But knowledge of the full set of requirements is in a much
more primitive state than for quantum mechanics, quantum field theory, or general relativity. These have given much understanding, and it
is likely the conformal group can also.

I.5 Transversions as point-dependent dilatations

There is an alternate view of special conformal transformations, other then regarding them as giving accelerations. They can be taken to perform dilatations, changing the (infinitesimal) distance between points, but with the dilatations functions of space (sec. II.2.d, p. 77; sec. II.3.g.ii, p. 93) [Barut and Haugen (1972,1973a,b); Kastrup (1966)].

This has been called a gauge transformation over Minkowski space, but is not the same as the gauge transformations of electromagnetism and gravitation (required properties of massless objects, and possible only for them [Kupersztych (1976); Mirman (1995c), sec. 3.4, p. 43]). The term is also used in other senses [Mirman (1999), sec. V.4.b, p. 261] so it is important to be careful of the name which can be misleading.

I.5.a Why does geometry require that physics be invariant under transversions?

It is this property of geometry, invariance under dilatations, both global and space-dependent, that imposes a requirement that physics be invariant under the conformal group. An observer measures the distance between two points (of course with a physical object), then moves to another position and measures distance again, with the same object, and compares distances. Can we say that lengths of instruments are unchanged by translations? The only way is by comparing them with an arbitrary length — another physical object. But this also must be translated. So if lengths of all physical objects are changed by the same factor in a translation it would be impossible to tell, there would be no experiment that could show it. Of course if there were an intrinsic length in geometry, that could be used. It is the lack of such a length that gives dilatation invariance. And if the invariance is global, how can it not also be local, and conversely?

Physical laws must thus be such as to be invariant these space-dependent dilatations otherwise they would make unverifiable predictions, they would distinguish lengths or points that cannot be distinguished. Another way of putting this is that there is freedom to pick physical laws that have (but not ones that do not have) this invariance, there is no experiment that can contradict such laws.

This does not guarantee that such freedom leads to new physical consequences (or does it?), but being present it might, so investigating it is important. One purpose of the present discussion is to raise this possibility and to start a study of some questions thus motivated.

I.5.b Why are these point-dependent dilatations and not accelerations?

How does this approach differ from that interpreting transversions as giving accelerations?

What is a rotation or boost? These are not transformations of an object to a different orientation or velocity, for that would require an acceleration, thus a force, and a nonconstant one, which the Lorentz group does not supply. Rather they are correlations between objects and equivalent ones that are rotated or boosted — operators of rotations transform statefunctions of objects into those that describe rotated objects (but the objects themselves are not rotated).

Likewise transversions correlate objects at rest or moving with constant velocity and ones with uniform acceleration, although the correlation is of somewhat different form. These transform statefunctions into ones describing accelerated objects, but in a distinct sense. For the acceleration interpretation, statefunctions on which we build representations (sec. I.3.d, p. 30) are different than ones for constant velocity. Transversions take paths of constant velocity and change them to paths of constant acceleration. Thus the set of basis vectors that we have to use are different, and these describe objects with constant acceleration.

However here we regard transversions as acting not on statefunctions but on the coordinate system transforming it to a another, equivalent, one. The observed behavior of objects is not changed, so their statefunctions are not. These are changed to different functions, and the variables, the coordinates, are also changed, so the transformed basis vectors are the same functions of the transformed coordinates as the untransformed ones are of the original variables.

The interpretations of transversions differ because the interpretations of what they act on do.

I.5.c Transversions are dilatations that vary with space

In what sense do transversions act as dilatations, and how do they vary (sec. II.3.g.ii, p. 93)? Intervals can be correlated, specifically small ones, and a change of size by a transformation acts as a dilatation. Thus we have to consider how the size of a small interval of transformed coordinates depends on the size of the untransformed interval, and whether this varies with space.

To show that transversions do give point-dependent dilatations we use (eq. III.1.c-3- III.1.c-4, p. 108; eq. III.1.d.vi-20, p. 114),

$$x'_\mu = \sigma(x)^{-1}(x_\mu + c_\mu x^2), \qquad (I.5.c\text{-}1)$$

$$\sigma(x) = 1 + 2c_\mu x_\mu + c^2 x^2. \qquad (I.5.c\text{-}2)$$

Then

$$\frac{dx'_\mu}{dx_\rho} = \sigma(x)^{-1}(\delta_{\mu\rho} + 2c_\mu x_\rho) - \sigma(x)^{-2}(x_\mu + c_\mu x^2)(2c_\rho + 2c^2 x_\rho)$$

$$= \sigma(x)^{-2}\{(\delta_{\mu\rho} + 2c_\mu x_\rho)(1 + 2c_\eta x_\eta + c^2 x^2)$$
$$-2(x_\mu + c_\mu x^2)(c_\rho + c^2 x_\rho)\}$$
$$= \sigma(x)^{-2}\{\delta_{\mu\rho}(1 + 2c_\eta x_\eta + c^2 x^2) + 2c_\mu x_\rho(1 + 2c_\eta x_\eta)$$
$$-2c_\rho x_\mu - 2c^2 x_\rho x_\mu - c_\rho c_\mu x^2\}; \quad \text{(I.5.c–3)}$$

so the dilatation is position dependent, and since the origin is arbitrary, its dependence is not fixed.

We can also use

$$x = \frac{aw + b}{cw + d}, \quad \text{(I.5.c–4)}$$

and

$$\frac{dx}{dw} = \frac{a}{(cw + d)} - \frac{c(aw + b)}{(cw + d)^2} = \frac{ad - bc}{(cw + d)^2}. \quad \text{(I.5.c–5)}$$

Thus $\frac{dx}{dw}$ varies along the path. The transformation is a point-dependent dilatation.

I.5.d   Invariance of angles makes the conformal group the largest physical group

If we can consider space-dependent dilatations why not more general transformations, larger groups than the conformal? The conformal group is the largest group under which angles are invariant [Porteous (1995), p. 245; Schottenloher (1997), p. 6]. And values of angles do not depend on an arbitrary physical unit. Suppose that we have an object defining angles (which are defined in a plane, and for the conformal group in general spaces there are such angles in every plane), say between two physical lines, or a closed figure of straight physical lines with the sum of its angles determined by the number of sides (a triangle for example), or an object whose behavior is governed by the projection of its statefunction onto a basis state (that is by the angle between the statefunction and the basis state). If we take it and rotate, boost, translate it, or do any other thing that does not change its properties, then the angles it defines are not changed. The sum of the angles of a triangle is invariant under these operations. And we cannot change other objects to keep angles the same, as we can change the mass of an arbitrary unit to keep all other masses the same, since physical masses are ratios. Angles are independent of measuring instruments.

Thus it is impossible to require that physics be invariant under change of angles, since it clearly is not, nor is geometry. But it is under the conformal group, with transformations properly defined. The conformal group is therefore the invariance group of geometry, thus presumably of physics, and is the largest such group. It is this that strongly implies its importance.

I.5.e   Transversions are unobservable

Is geometry, thus physics, really invariant under transversions — dilatations that are functions of space? Could we not tell if such transformations are carried out? We need observers who look at themselves — at objects moving with them — and also at other objects. Is there a way for them to determine a difference between before and after such a transformation?

I.5.e.i   *The reference curve*

Pick an arbitrary curve to serve as a reference curve. From each point on it draw a line to a point on a second curve, the path of an object, with the angles between these curves and the connecting lines the same for all points (which is possible since angles are not changed by transformations, like translations), thus correlating points on the two curves. Numbering points on the reference curve thus gives the numbering for the path. Since points are dense this cannot actually be carried out, except for a countable subset.

How are lines drawn that are at fixed angles to both curves? Of course this can only be done with some curves, but including those we wish to consider, paths of objects with constant velocity, straight lines. A reference observer can be a system that emits say photons a pair at a time (in principle), at equal time intervals, and all with the same directions. One photon strikes a film on the reference curve, marking it, the other hits a film on the second curve, the path, also producing marks. If the curves are of constant velocity then the angles between the paths of photons and the curves and distances between marks are constant.

However we need only assume that distances between tick marks on each curve are equal, as measured by the observer on that curve, and also that time intervals are equal. Since we know that it is possible to have observers moving with constant velocity, and different velocities, this is always possible (and if it were not possible then there would be enough freedom in definitions to take it as correct). Each observer lays a physical object, a measuring stick, between ticks on its curve, and finds that it always just fits, no matter what pair of ticks is chosen. And a clock that moves with the observer always records the same interval of time for an object with constant velocity to move between two marks.

I.5.e.ii   *How transversions affect the curves*

Let us now carry out a transversion. What is the difference in geometry and physics before and after a conformal transformation? Angles
between those lines correlating points and the two curves are invariant
under conformal transformations. And all distances, lengths of physical objects, and time intervals, change by the same (space-dependent)
factor. Thus each observer, both before and after, sees its measuring
rod just fit between ticks, and the time intervals marked by the clocks
the same — it measures the same distances and time intervals.

Moreover, either before or after a transformation, when it encoun-
ters a tick it sends to the other observer a signal along the line correlating the points (taking the observer as a long object). Angles are
unchanged so signals reach the second observer at the corresponding
tick marks on its curve, and at equal time intervals as measured by that
observer. This one looking at the other curve finds the lengths along it
remaining equal, and the time intervals for covering these equal lengths
also the same. Neither observer, looking at its own path, or that of the
other, can find a difference due to the transformation. It is physically
unobservable.

Which set of distances is correct, the one assigned before a transformation, or that after? Physically (and geometrically) there is no way
of knowing. We could have started with the second set and performed
the inverse transformation, getting the original set. It is impossible to
decide which is preferable — they are completely equivalent.

I.5.e.iii   *Is there ambiguity in measurement of time?*

Some of these concepts can be illustrated (in a way that does not exactly correspond to the concepts discussed here) by considering how to
measure time. We are familiar with clocks, say given by frequencies of
agreed-upon spectral lines. But might there be some physical freedom,
or physical preference, in deciding which clocks we should use [Mirman
(1995c), sec. 10.4, p. 179]?

We might choose the expansion of the universe as the most fundamental clock. We look at a distant galaxy and from the Doppler shift we
know its speed away from us. Let us assume that we know its intrinsic
brightness, and then using the apparent brightness we find the size of
the sphere over which its radiant energy is spread, so the distance from
us. Then we find the function of distance that the speed is, so how fast
the universe is expanding, thus when the distance was zero — the age
of the universe. Knowing the speed of light and the distance we can
find when the light was emitted, thus the age of the universe at that
time.

However there is a slight problem. While the light is in transit the
universe is expanding, thus the radius of the sphere we use to relate

the intrinsic and apparent brightnesses is larger than it was when the light was emitted — the apparent brightness is decreased since the light energy is spread over a larger area. The actual distance to the galaxy (when the light was emitted) is less than the distance calculated from the apparent brightness — we use the (incorrect) calculated distance and the (misleading) observed brightness to find the intrinsic brightness, which is less than the actual value.

If we assume that the speed of a galaxy, thus its Doppler shift, due to expansion increases with distance from us we find that the speed of the galaxy is less than given by its assumed distance (because the distance determining the Doppler shift is smaller than the assumed distance which is based on the apparent brightness). Or if the Doppler shift is used to find the distance, from the apparent brightness we determine an intrinsic brightness which is incorrectly dim, so the galaxy appears further away. Thus calculating the expansion of the universe from these distant galaxies, and from ones nearby, we find the rate of expansion obtained from the latter is greater than from the former so the expansion of the universe seems to be increasing, the universe is accelerating. (Of course we cannot conclude that the universe is accelerating without applying this correction.)

Suppose that when the light was emitted a clock was also (perhaps a cloud of radioactive atoms, so that we can find the time of emission from the ratios of elements). The clock travels with the speed of light (actually less of course, but its speed can be measured and accounted for). The time that it reads, the time it takes the light to travel to us, is less than that we calculated.

This can, in principle, be used to study the expansion of the universe.

Which time is the physically correct one, the calculated time, or that given by the clock? Are there criteria that we can use to decide, or can we just change definitions, of acceleration for example, to allow us to pick whichever time we wish?

This illustrates that there may be (physical) freedom in choosing coordinates.

I.5.e.iv *So physical laws must be invariant*

All lengths and times are transformed by the same factor (sec. I.2.a.i, p. 9), all masses by its inverse (sec. I.2.a.v, p. 12). Therefore physical laws are the same whether a transformation was carried out or not. This means that going from one point on the line to another, or from a point on the reference curve to one on the line, all physical objects are so transformed — their lengths by the same factor, their masses by the same inverse one. It is impossible to tell whether a transformation was carried out. This being so, physical laws must be invariant under these transformations otherwise they would distinguish coordinate systems that cannot be distinguished. This is the argument for invariance under the conformal group — including transversions.

I.5.e.v *The difference between transversions and arbitrary transformations*

Space-dependent dilatations are not arbitrary. There are free parameters, but once these are given, the functions of space that dilatations are become fixed since they are members of the conformal group.

If we can arbitrarily vary the shape of a curve using transversions, can we not use these to carry out arbitrary transformations, so that any transformations would be possible using the conformal group? Transversions, being conformal transformations, conserve angles, as arbitrary ones do not. Thus what is the difference between an equilateral hyperbola and an arbitrary curve? Can a transversion transform between these?

Consider tangents to these curves at the point where they cross the (arbitrary) z axis. The angles between the lines are different for the various curves. Thus transversions, which leave angles unchanged, cannot transform between them.

These space-dependent dilatations are unobservable, but would be observable if such angles changed. They cannot be recognized because transversions affect all objects in the same way, they multiply all physical and geometrical quantities of the same type by the same constant. Arbitrary transformations can take a circle into an ellipse, or an equilateral hyperbola into one that is not, or into an arbitrary curve, because, for example, they dilate lines at different angles by different amounts, so changing angles.

I.5.f Determining whether a line is straight

Straight lines are distinctive curves in Euclidean geometry (sec. I.2.c.ii, p. 18). But what is a straight line? Ones that are parallel never cross, but they do not cross after a conformal transformation; this leaves angles invariant, a reason that conformal groups are closely related to geometry, as others (except subgroups) are not. Also a straight line is the shortest distance between two points. But this requires a distance function. If that is changed, as under a conformal transformation, we could not tell. Thus we cannot pick the geometry before or after a transformation and say that it is the correct one. All (pseudo-)Euclidean geometries related by conformal transformations are equivalent.

Might an observer on a curved path know that it is accelerated, so know that its path is curved? In particular if charged particles do not radiate before a transformation (they were moving with constant velocity, so on a straight line), would they after? An observer would have to perform experiments that would show this. But the effect would be canceled if units changed. Observers on a curve would always see themselves at rest, both before and after a transformation. If a charged particle did not radiate in one case, the transformation could not change this as it does not introduce a force to accelerate the particle. (This is different from the correlation of statefunctions of accelerating and nonaccelerating objects.) Coordinates change but in such a way to leave velocity constant, otherwise the transformation would be noticeable, which it is not.

What would another observer, say the reference observer, see? This one marks off points on its own path at equal times, which since it is moving at constant velocity are at equal distances on the path. At the same set of times it sends signals to the other path, which intersect it so as to also mark off equal times, as seen by the reference observer. An object on the curve responds to these ticks by sending signals back along the path to the reference observer, which has a trailing part able to receive them. Thus this observer sees signals sent at equal times, and as signal paths are at constant angles to the curves, these signals appear to be sent at equal distances — it does not see the particle accelerated, whether a conformal transformation was carried out or not. This transformation does not change angles of the various paths.

Thus if an observer sees the velocity of another observer as constant, it will after a transformation, and conversely. With this interpretation of conformal transformations, transversions, as being changes of scale, a constant velocity does not become an acceleration. An observer that sees objects moving with constant velocity always sees them moving with constant velocity — conformal transformations do not change this.

I.5.g   What are the conformal group basis states?

To give a representation we specify, among other functions, its basis states. What are the proper basis states for the conformal group implied by these considerations? Here we outline requirements. For the rotation group there are sets of functions of the angles, for it spherical harmonics, the members of each set mixed among themselves by the group operations. These, or linear combinations of them, provide the rotation basis states. The conformal group is more complicated, as is even the Poincaré group. We have a set of conformal basis states (sec. I.3.d.iv, p. 34) but these are for constant acceleration thus for two dimensions. What we need are states for any dimension, specifically 3+1. These would be states mixed among themselves by the group transformations. One subject to be studied is the set of (hyper)surfaces invariant under these transformations. For the rotation, Lorentz and translation groups, these are circles, spheres, hyperbolas, cylindrical hyperbolas, lines, planes, and generalizations to higher dimensional spaces, for the conformal group sets of these. Since we considered only two dimensions we obtained hyperbolas, or for a plane immersed in a larger space, cylindrical hyperbolas. But we also wish to study transversions over spaces larger than the plane. These would leave invariant sets of hypersurfaces which we need characterize. Here Clifford algebras (sec. II.5, p. 98) are likely to be useful. The basis states would then be functions varying only over such surfaces, as the rotation group states are functions on spheres, leaving each sphere (radius and center) fixed.

But this raises several questions. Finding such states will probably not be a major problem, finding all, and showing that they are all, might be more difficult. There are basic questions. What do we mean by a representation, by basis states? What criteria are there for picking these? For the rotation group, as ordinarily realized, and this is not the only realization (sec. III.3.a, p. 120), there are definitive sets of functions that satisfy, that are mixed among themselves, and that are all such sets. Any (well-behaved) functions of angles can be expanded in terms of them. But for more complicated groups, especially nonlinearly-realized ones, like the conformal group, even simple ones, it is not clear that there are distinct sets of functions like these, or that they are complete. The (required) nonlinear realization of the Poincaré group for gravitation [Mirman (1995c), sec. 11.3, p. 190] gives at least a warning. Moreover while for the Poincaré group and its subgroups we find representation basis states by finding functions defined over (hyper)surfaces, there seems to be no proof that this will (always and completely) work for other groups.

One way of approaching this is to find the (hyper)surfaces that are invariant under inversions generating the group (sec. I.3.e, p. 35). For two dimensions they are unit circles, or equilateral hyperbolas, for definite and indefinite metric spaces. Inversions in these leaves them invariant, and takes every parallel circle (or hyperbola) into another that is also parallel, so the set of these is invariant. The points of a sphere are labeled by spherical coordinates and for other spaces (particularly of 4 dimensions) by their generalizations to hyperspherical coordinates. To go to a space of dimension 3+1 the proper set of sines and cosines is replaced by the hyperbolic equivalents. This then gives the set of functions invariant under inversions giving transversions. As with two dimensions, exponentials in these functions would give basis states. However it is not clear that this procedure would give all types of representations, beyond those that can be expanded in terms of these functions. Also we can use generalizations of spherical harmonics as basis states, and these two types could (probably) be expanded in terms of each other (sec. A.6, p. 240). Whether these form sets, each a different representation, has to be studied.

Finding (all) relevant functions for the conformal group, and their properties, will likely be an extended — but useful and interesting — task.

I.6   Is the singularity of the transformation relevant?

Transversions have singularities as can be seen for the complex plane (sec. II.2.a, p. 68; fig. II.2.a.vii-2, p. 73; sec. II.3.f, p. 89; fig. II.3.g.iii-1, p. 95). In general there is a point on one curve with coordinate z_s such that (eq. I.1.e-4, p. 8)

$$\sigma(z_s) = 0, \tag{I.6-1}$$

$$\sigma(x) = 1 + 2c_x x - 2c_t t + c^2(x^2 - t^2) = 0, \tag{I.6-2}$$

at every point (at every value of t). On the curve then both

$$x' = \infty \text{ and } t' = \infty, \tag{I.6-3}$$

and these points are correlated with points on the reference curve for which neither coordinate is infinite. Thus an object traveling along this line would reach what it measures as infinity at a point that corresponds to a finite one on the reference curve. However it would take an infinite amount of time to do so. This should not lead to unphysical effects. The reference observer should not notice anything strange since it sees the second object reach infinity after an infinite time as measured by the object's clock.

I.6.a   Can points at finite and infinite distances from the origin be correlated?

Is there anything wrong with a point at a finite distance from the origin on one curve being correlated with the point at infinity on another one? Consider two curves, ignoring conformal transformations. Points are correlated, perhaps by an observer that consists of a set of atoms emitting light, one striking one curve — its coordinate line — which has a film so giving dots marking the line, the other striking a similar film on the second one, thus correlating the two lines. Points marked this way correspond. Are these the correct ones to correspond? One observer might ask the other to measure its distance from the zero point on that curve and require that the distances between zero points to corresponding marked points be the same on the two curves. But how do observers know that their units are the same? It is possible to take a unit length for one curve and move it to the other to compare. But this might change the size of the unit (as conformal transformations can be considered to do). There is thus no way that units can be determined equal, except arbitrarily. Observers pick pairs of points to correspond and the distances between them are the unit lengths for the two curves. But if all lengths (and correspondingly time and inverse mass) were stretched by the same factor on one curve, the unit-length objects for the two curves would still be the same — their end points would still correspond.

Since an observer cannot match points except arbitrarily it is quite reasonable that it correlates a finite point with that at infinity on another curve. It is well-known that there are one-to-one correspondences between points on a line and points on segments of it.

One observer might take the correlation lines as perpendicular to its own axis. However it cannot require or ensure that these are perpendicular to the other curve (but must rely on the other observer). The path can be for a different velocity or curved — there is no way of knowing. There is no natural way of relating points on two curves.

It is not possible to know whether correlations between points at infinity with finite reference points (and conversely) are the result of a conformal transformation or of the way points are correlated. Special conformal transformations have no effect that can be experimentally determined, on geometry or physics, so physical laws must be invariant under them, thus under the entire conformal group.

I.6.b   The conformal group and causality

Transversions can change the sign of the Poincaré-invariant square of the interval between two points, thus interchanging space-like and time-like intervals. These interchange the inside and outside of the light cone. Does that lead to violation of causality [Carruthers (1971); Laue (1972); Rosen (1968); Wess (1960)]?

Since (eq. I.1.e–4, p. 8)

$$(x' - y')^2 = \frac{(x - y)^2}{\sigma(x)\sigma(y)}, \tag{I.6.b–1}$$

and there are points for which $\sigma(x)$ is zero, so passing through such points changes the sign of $(x - y)^2$, thus interchanges space-like and time-like intervals.

Transversions map light cones into light cones — however they map finite points of light cones into finite points of light cones. This is not true for points at infinity [Wess (1960)] which can be mapped to finite points not on light cones, and conversely.

Is interchange of the inside and outside of the light cone, an apparent violation of causality, physically relevant? It would be only if it were necessary to carry out these transformations. But are we required to, can we?

I.6.b.i *What transformations are physically necessary?*

Consider the rotation group. For any observer there is another (possible) one at every angle. There is no way of choosing one over another — they are physically equivalent, distinct but not different. All directions in space are equivalent; space is rotationally invariant. Yet even if were not so invariant it would still be possible to have observers at every angle. Thus the vertical seems different from other directions. And with a magnetic field the direction in which it points appears distinct. Yet it is possible to have angular momentum point in any direction, along the vertical or at any angle to it, or to a field. Angular momentum may not be conserved in such cases, it may vary in time, but there can always be an angular momentum pointing anywhere. Thus observers (coordinate systems, angular momenta) can have any orientation, so it has to be possible to transform observations of one to those of any other. We therefore must be able to carry out any rotation. And if we perform a set of rotations, we also can perform their products since all products, no matter how many, no matter in what order, take one possible observer to another.

Thus all possible, so necessary, physical rotations, are in one-to-one correspondence with the elements of the rotation group. Also the rotation group operators, and representations, are analytic functions of angles, always well-behaved.

I.6.b.ii *Not all transversions need be carried out*

But consider transversions. We have argued that such space-dependent dilatations cannot have experimental effects, we cannot determine if one has been carried out or not. Thus physical laws must be invariant under them otherwise they would make predictions that differ from what is observed. But this physical irrelevance is not true if a transformation is singular, or reverses the inside and outside of the light cone, that is violates causality. We cannot require that physical laws be invariant under such transformations.

Thus whether a transversion is a symmetry of physical laws depends on its parameters, unlike the rotation or Lorentz groups, say. And if we can require physical laws be invariant under two transversions, we may not necessarily be able to require them to be invariant under their product (and they might not be). If transformations can have no observable consequences, physical laws must be invariant under them, but not otherwise (sec. IV.3.b.i, p. 199).

Transversion transformations of the conformal group are different from transformations that may be more familiar. There are some that are singular. This is not true of the Poincaré group or its subgroups, finite or continuous. Nor is it true of SU(2,2) or SO(4,2) or SO(5,1), as ordinarily realized (sec. III.2.b.ii, p. 120). But when these groups are realized as transformations over a space of four real dimensions, smaller than the spaces over which they are defined, giving the conformal group, some of their transformations must be nonlinear. And so they can, and in this case clearly must, have singularities, and other strange properties. These arise from nonlinearities, and the type of realization used.

I.6.b.iii   *The conformal group does not violate causality*

Hence there is no objection to some conformal transformations violating causality. Usually we regard the set of physical transformations to be in one-to-one (or perhaps two-to-one) correspondence with the group elements — the set of physical transformations and the set of group elements form homomorphic sets. It is this that fails here. The set of physical transformations is smaller than those of the group.

An illustration is given by a flat surface that ends abruptly, such as a table. That the effective laws of physics are invariant under two translations (or rotations) does not imply that they are invariant under their product. And whether they are invariant under a transformation depends on its parameters and the initial position. Effective physical laws must be such as to be invariant under transformations that take one point on the table to another point on the table, otherwise they would distinguish two points that cannot be so distinguished. However this does not mean that they are invariant under the full inhomogeneous rotation group, ISO(2). Clearly they cannot be.

Group transformations that seem objectionable do not really cause (physical) problems because they do not correspond to physical transformations. In general it is necessary to be careful in identifying (all) group transformations with physical ones [Mirman (1995c), sec. 3.4.3, p. 46].

The set of transversions under which physical laws are required to be invariant has to be determined in specific contexts. However there is such a set and here we explore only the consequences of that; we do not need to find its range.

Again we see that the rotation group, and also (but perhaps less so) the Lorentz and Poincaré groups [Mirman (1995c)] are quite simple and in some ways special, so care is required when generalizing what we learn from them to cases that might be very different, as for example with nonlinear operators (clearly something of great importance in physics) and unusual spaces of definition, and likely in other ways.

I.7   The conformal group and interactions

Among the most fundamental facts of nature are that objects interact, they obey nonlinear equations. They must, otherwise we would not know of them and they would not exist (nor would we). But why do they interact, why are the equations nonlinear? What requires nonlinearity, as presumably something does? It is known that the conformal group implies the presence of interactions. Properties of the group suggest that it might provide hints, but unfortunately at this stage nothing more; these can at least stimulate thinking. Of the known interactions, the ones to which the hints seem relevant, which perhaps help us understand these hints, are the electromagnetic and gravitational.

We review their relevant properties [Mirman (1995c)] and consider analogies suggested by the conformal group. In particular, as we see, the field given by transversions has a self-interaction, like the gravitational field. Yet what is the effect of physical objects on it? For suggestions we have to study the gravitational field.

I.7.a Do physical objects have to produce gravitational fields?

It is clear that gravitation must be nonlinear, so the nature of free gravitation is determined [Mirman (1995c), chap. 8, p. 135]. And the behavior of objects in a gravitational field is also determined by the transformation properties of the functions describing them, like momentum [Mirman (1995c), chap. 5, p. 73]. But what determines how these objects affect the field, or even if they should? Why, and how, do physical objects produce gravitational fields?

One argument is that a field cannot affect an object without the object affecting the field. However this is a philosophical prejudice that might not be shared by gravitation. What is intrinsically wrong with space being curved? Then the path of an object would be determined by the curvature of space, but that could be independent of the body. Momentum would not be conserved, but it should not be if space were not homogeneous.

What this belief really says is that without objects space is flat, homogeneous, isotropic. While there seems no (known) reason to insist on that, it is the simplest assumption, and quite sensible. Let us therefore assume that it is true. Thus the presence of objects must produce a gravitational field. How?

To the gravitational momentum operator we add a term nonlinear in statefunctions of objects [Mirman (1995c), chap. 9, p. 150], called the energy-momentum tensor (although it is not really a tensor [Mirman (1995c), sec. 9.2.1, p. 154]). What is the form of this? While there seems now no known requirement that determines its form we can place reasonable conditions. It must be a two-indexed object, $T_{\mu\nu}$, and it must be complex.

Gravitational momentum operators are the covariant derivatives, (schematically)

$$\Gamma_{;\mu} = \Gamma_{,\mu} + \ldots, \tag{I.7.a-1}$$

and

$$\Gamma_{,\mu} = i\frac{d\Gamma}{dx_\mu},$$ (I.7.a-2)

with the "i" necessary [Mirman (1995b), chap. 2, p. 25]. Hence Γ is a complex function. Of course Γ is a wave, so of the form of integrals of $exp(ikx)$, and complex, as is well-known [Misner, Thorne and Wheeler (1973), p. 945; Schutz (1993), p. 230; Weinberg (1972), p. 255]. The metric, the connection, so functions of these, curvature and the Ricci tensor, are all complex. So T has to be complex.

There are accepted expressions for T [Mirman (1995c), sec. 9.2.2, p. 155]. For a scalar field, with

$$\phi \sim \int a(q)exp(iqx)dq,$$ (I.7.a-3)

integrated over q,

$$T_{\mu\nu} \sim \phi_{,\mu}\phi_{,\nu} + \ldots,$$ (I.7.a-4)

and is properly complex. For a spinor, T is taken as

$$T_{\mu\nu} \sim \psi^+\gamma_\mu\psi_{,\nu} + \ldots,$$ (I.7.a-5)

the product of a spinor and its conjugate. Now ψ is a sum (actually an integral) of the form

$$\psi \sim a(q_1)exp(iq_1x) + a(q_2)exp(iq_2x) + \ldots,$$ (I.7.a-6)

where the a's are spinors and complex since different terms have different phases. Thus

$$T_{\mu\nu} \sim q_1|a(q_1)| + q_2|a(q_2)| + \ldots$$
$$+q_1a(q_1)^+a(q_2)exp\{(i(q_1 - q_2)x\} + \ldots;$$ (I.7.a-7)

again properly complex.

If we assume that energy-momentum tensors are of the lowest order in statefunctions they seem then to be fixed. This illustrates the argument, but since it involves several assumptions, that are quite reasonable but not definitely required as far as presently known, it does not seem worth checking if there is still some slight freedom in choices of couplings of gravitational fields to matter, except of course for the quite mysterious value of the coupling constant.

This at least suggests questions that might be thought about in trying to find why the couplings must have these forms, as presumably they must.

I.7.b   Nonlinear momentum operators

Realizations of transversion Lie algebra operators, the K's, are nonlinear (sec. III.1.e, p. 115). However nonlinearities — interactions — that are fundamental in physics are of a different type. Does the conformal group lead to these, or at least suggest ways in which they should occur, and why?

Lie algebra operators are differential operators in coordinate variables, x's, and the K's are nonlinear in the x's. Thus (always schematically),

$$K \Rightarrow ix^2 \frac{d}{dx}. \qquad (I.7.b-1)$$

These, as well as the other algebra operators of the group, act on functions ψ that are solutions of the differential equations so obtained (from the group invariants and the labeling operators), and are representation basis vectors.

However momentum operators, which contain the interactions, are nonlinear, not in the x's, but the ψ's. That is

$$p \Rightarrow i\frac{d}{dx} + \psi, \qquad (I.7.b-2)$$

so

$$p\psi \Rightarrow i\frac{d\psi}{dx} + \psi^2. \qquad (I.7.b-3)$$

Momentum operators depend on the solutions, and do not merely act on them. How can this occur?

Gravitation shows how such interactions can arise [Mirman (1995c), chap. 8, p. 135]. A statefunction of an object is transformed by a Lie group transformation,

$$\psi \Rightarrow U(x)^{-1}\psi U(x), \qquad (I.7.b-4)$$

where U cannot be written in the form, with R the Lie algebra operator,

$$U \sim exp(iRx), \qquad (I.7.b-5)$$

because it is nonlinear in a gravitational field. Being in such a field, it is a function of position.

I.7.b.i   *Why momentum operators are covariant derivatives so contain interactions*

The momentum operators of the Poincaré group are vectors by the (required) definition of the group. But the ordinary derivative $\psi_{,x}$ is not a vector since U is a function of x, so $(U(x)^{-1}\psi U(x))_{,x}$ contains derivatives of U. Thus extra terms must be added to the momentum operators to cancel these derivatives, and the momentum operators are then covariant derivatives. That is why covariant derivatives are required. These extra terms are functions of the connection, Γ — they are nonlinear in Γ.

Now $\Gamma^{\alpha}_{\beta\gamma}(x)$ changes under Lorentz transformations; different observer of course see the same Γ, but the components are different in different systems. And

$$\Gamma(x) = T(x)\Gamma(0), \tag{I.7.b.i-1}$$

where T is the translation operator, again not an exponential because of nonlinearities. Hence Γ must be a basis vector of the Poincaré group. Momentum operators are thus needed that act on it. However these cannot be ordinary derivatives, which do not transform properly, but rather covariant derivatives, $\Gamma_{;x}$, found by adding to the ordinary derivative terms to cancel the derivatives coming from the transformations, in the well-known manner [Mirman (1995c), sec. 1.2.2, p. 8]. The extra terms are functions of Γ, as is true both for the statefunctions of massive objects, and for Γ itself. Thus (schematically)

$$p_x \Gamma = \Gamma_{;x} = \Gamma_{,x} + \Gamma\,\Gamma. \tag{I.7.b.i-2}$$

The gravitational field interacts with itself (for this reason); momentum operators are nonlinear, and are nonlinear functions, not of variable x as with K, but of statefunctions, here Γ.

I.7.b.ii *Interactions given by the conformal group*

With this model, what can we say about interactions in the conformal group? Write as $V(x)$ the group operator for a transversion, a function of x because these can be taken as arbitrary (to some degree) space-dependent transformations, space-dependent redefinitions of coordinates, and again because of nonlinearity not expressible as exponentials. Then as with gravitation

$$\psi \Rightarrow V(x)^{-1}\psi V(x), \tag{I.7.b.ii-1}$$

so momentum operators cannot be ordinary derivatives, but must be covariant ones. To ordinary derivatives must be added terms, denoted by κ, to cancel the derivatives of V.

By the same argument as for gravitation, terms so added must themselves be Poincaré representation basis vectors. The momentum operators that have to be defined to act on them are covariant derivatives, containing κ itself. So

$$p_x \kappa = \kappa_{;x} = \kappa_{,x} + \kappa\kappa. \tag{I.7.b.ii-2}$$

Hence this field has an interaction, like gravitation, a self-interaction.

II.1.c Probability and the absolute square of statefunctions

Probability is given by the absolute square of statefunctions. Why? There are two questions here (beyond the already discussed need for probability [Mirman (1995b), sec. 3.3, p. 47]), why the absolute value, and why the square? Are these just incomprehensible postulates that just accidentally happen to be true, or can we derive them, understand them, from more basic requirements?

Probability usually refers to the "probability of a particular classical outcome", that is if we perform experiments and find a mark on a film, a bubble in a chamber, a pointer, say, all in a certain position, the probability (distribution) is the function of position that gives the fraction of the experiments in which the macroscopic object is found at that position (or within a small interval about it). We use experiments such as these to illustrate the arguments. The term probability might appear in other ways (so for example in interference experiments, if we wish to discuss probability for them, phases are quite relevant). But the use here is the most common, and consideration of why any aspects of these arguments are irrelevant for other cases aids understanding of them. Words are often used in many ways, so it is impossible to discuss all types of experiments in which a word like probability might be used. However these illustrations should be sufficient to show why such rules of quantum mechanics are necessary, and how they arise.

II.1.c.i *The phase cannot appear*

There are several reasons why the phase of the state of the incoming object (of the cloud chamber experiment, for example [Mirman (1995b), sec. 3.2.1, p. 43], but the considerations are general) is irrelevant, so only absolute values give observed results. The initial value of the phase is arbitrary, for example when it depends on the origin of the coordinates, as with $exp(ikx)$ or $exp(i\omega t)$, and this origin is arbitrary. Also to determine statefunctions we must repeat experiments many times, and phases average to zero.

Also what is relevant is not the phase (the phase of a single object cannot be measured, so has no meaning [Mirman (1970), p. 3356]), but the phase difference between the incoming object and, say, the struck molecule. But we do not measure the latter, so neither phase can affect observations. And if we did measure the phase of the molecule, we would change it. The statefunction of the atom resulting from the breakup, so of the system which the bubble in the chamber or the mark on a film is for example, does depend on this phase difference. But even if we knew it, the dark spot is altered by random motions of objects in its environment, and these average out phases of statefunctions of the spot. Also the phase of a statefunction of this dot determines details

I.7.c.ii *What are the possible forms of the transformations?*

This can be considered another way, using, say, Dirac's equation (generally the two equations of the Poincaré-representation labeling operators). It is invariant under Poincaré transformations. What other transformations can we use? One is the dilatation, under which it is also invariant (sec. I.2.a.viii, p. 14). Suppose that we consider space-dependent transformations. Then

$$\psi \Rightarrow U(x)\psi. \tag{I.7.c.ii-1}$$

One possibility is that U multiplies both components of the spinor by the same number. Since we keep the normalization constant (for if the number of objects changed under the transformation it would be very unlikely to give invariance) then U can only be a phase. So

$$i\gamma_\mu \frac{d\psi}{dx_\mu} \Rightarrow i\gamma_\mu \frac{d\psi}{dx_\mu} + i\gamma_\mu \frac{dU}{dx_\mu}, \tag{I.7.c.ii-2}$$

which for invariance of the equation requires another term, with one index, which is also changed by the transformation by the addition of a term to cancel $i\gamma_\mu \frac{dU}{dx_\mu}$. This is the electromagnetic potential. The other possibility is that U mixes components, and by the same argument there has to be an extra term in the equation, but now with three indices, the gravitational field.

This is the case in which the transformation is a space-dependent Lorentz transformation. Apparently this is all that we can do. More general transformations, say those acting on components differently, perhaps without mixing, do not keep $i\gamma_\mu \frac{d\psi}{dx_\mu}$ unchanged, so there seems no way that Dirac's equation would be invariant, which is what we require (or believe in).

I.7.c.iii *Why are these fields connections?*

Why are the electromagnetic potential and the gravitational field, Γ, connections? These two are massless objects, and are connections since their little groups [Mirman (1999), sec. VI.2.c, p. 284] are solvable, rather than simple [Mirman (1995c), sec. 2.4.2, p. 22]. And because of the difference of the types of little groups, for gravitation to couple to massive bodies it must satisfy a nonlinear relationship, the Bianchi identity [Mirman (1995c), sec. 8.1.2, p. 138].

A connection is an object that under a transformation may have its components mixed like a tensor, but also terms added. The elements of representation matrices of a solvable group are all zeros on one side of the diagonal, unlike those of semisimple groups for which elements on the two sides are related, requiring nonzero elements on both [Mirman (1995c), sec. 3.2, p. 33]. When acting on an object (to transform it) such matrices give added terms, while representation matrices of semisimple groups cannot.

I.7.c.iv *How the Poincaré group gives connections*

To find representations of the Poincaré group we use little groups [Mirman (1995c), sec. 2.2, p. 12], taken as ones leaving a momentum invariant, $p = (1,0,0,0)$ for massless representations, and $p = (0,0,0,0)$ for massive ones. We cannot use the latter for massless objects since they can never be at rest, so there is no Poincaré representation for them with that vector. But we use it for the massive objects since that is the largest possible subgroup, SO(3), that we can take (and with known representations) giving the smallest number of other Poincaré generators remaining to be determined. For massless objects then the little group from which we built the representations is SE(2), which leads to gauge transformations, and is solvable. Thus there are Poincaré representations for which some generators add terms; these are connections. That is why connections occur.

I.7.c.v *The analysis for transversions*

For transversions the analysis is similar to that for gravitation. Statefunctions undergo space-dependent transformations, so (for example in Dirac's equation, with the analysis similar for the Poincaré invariants which give the conditions for all spins [Mirman (1995b), sec. 3.4.4, p. 56]) to be able to cancel the derivative of a transformation we must insert a three-indexed object, Δ, a connection. The effect of a transversion can be considered to be an acceleration; its action on a statefunction is independent of its interpretation. It is thus the same for the interpretation of it being a space-dependent rescaling. But the action of an acceleration is the same as that of a gravitational field, and this changes the direction of vectors, mixing components. Also a transversion can be taken as the product of a translation with an inversion of the unit sphere and then another translation (sec. III.1.d, p. 108). An inversion changes directions of vectors; a transversion contains an inversion, thus the effect is to change directions of vectors, mixing components.

Another way of looking at this is that transversions are generated by Lie algebra operator K_μ acting on a tensor (statefunction), $K_\mu T_\rho$, where ρ runs over the states of the Lorentz group representation to which T belongs. Considering the more familiar rotation subgroup, we have a product of two representations (one, the K, being the vector representation) decomposed into a sum [Mirman (1995a), sec. XII.2, p. 340] out of which is projected the representation to which T belongs since the transformed statefunction has to be of the same representation because the transformation leaves the physical situation — the experimental results — invariant, and generally by the definition of representations.

However the state label, the equivalent of the "z-component of the angular momentum" is changed by the multiplication by K. Thus, for example, the direction of a vector is changed, so its components are mixed.

I.7.c.vi *What is the little group, and why?*

For the conformal group the little group is the Poincaré group; we take its representations and build those of the conformal group from them. However this is neither semisimple nor solvable, but inhomogeneous. But we need, for equations to be invariant, a solvable group whose transformations add a term to Δ. Here we have more freedom than for massless or massive cases to pick the little group upon which to build representations since the requirement of simplicity does not place restrictions on the choice of the little group, as for the Poincaré case.

Thus we chose the type of Poincaré representation by building it on a subgroup, the proper one is solvable, and this the Poincaré group has [Mirman (2001), sec. I.7.b.iii, p. 39]. That is we construct the representation according to which Δ transforms from those of the Poincaré group. But this has different types of representations [Mirman (1995c), sec. 1.1.1, p. 2]. The proper type is that which gives Δ as a connection, so the point-dependent transformations that change statefunctions of other objects can also add terms to Δ to cancel these changes and leave momentum operators, so Dirac's equation say, invariant. These representations are the ones giving such invariance, so are the ones we must choose.

A field resulting from the transversions is a connection, like electromagnetic potentials and gravitational fields.

Since these transversion transformations cannot (now) be related to known physical fields, it does not seem to be of great use to consider details.

I.7.d   How these relate to realistic interactions

Thus we have considered three interactions, the electromagnetic, the gravitational and that suggested by transversions. Of these the one that has the strongest rationale, the transversion one, is also the one that seems to have no relationship to nature — unless we can think of a reason to identify it with gravitation. It is in addition interesting that transversion generators, K's, can be realized with parts acting on internal variables (sec. III.1.e, p. 115). Angular momentum operators of the group can also be so realized, with the internal part acting on spin. The K's contain these spin operators, and in addition, there is another operator, a Lorentz scalar, that also appears in dilatations, and in the K's these two are space-dependent. The third internal operator is independent of space. From the K's we form group transformations, and these then have space dependence. This suggests that if we can find a reason to identify part of these transformations with gravitation, we might be able to identify another part with electromagnetism. At this stage however whether this is relevant is quite unclear.

I.7.d.i   *Why there are electromagnetic and gravitational interactions*

The electromagnetic and gravitational interactions are necessary for invariance because we can perform space-dependent transformations on statefunctions of other objects. But the reason that we can perform such transformations is that these fields exist, and are coupled to other objects, and are mass-zero representation basis vectors, so connections, allowing addition of terms that cancel those from transformations on statefunctions. Their properties (such as the required linearity of electromagnetism and the nonlinearity of gravitation, and that there are only two massless fields, are fixed by the representations under which they transform, and that they couple to massive objects [Mirman (1995c)]. Given their existence, their representations and that they so interact, they, and the form of their interactions and that massive statefunctions can undergo arbitrary space-dependent transformations, are essentially determined. But we have no clue (now) why they do exist, and why they do interact.

For transversions we can argue that geometry must be invariant under their transformations, leading to interactions with the resultant field. Here then the problem is just the opposite. We know why such a field exists, and why and how it interacts, but have no clue why it does not seem to appear physically (unless perhaps it does, but we do not recognize it).

These may seem similar to Weyl's gauge theory [O'Raifeartaigh and Straumann (2000)] but the physical and geometrical foundations are different.

I.7.d.ii   *Interactions with mesons are different*

Gravitational and electromagnetic interactions are not the only ones. Given that, say, nucleons and pions exist, and their properties, why do they interact? Pions do not (apparently) allow transformations on nucleons that leave laws of physics invariant but which would not if they did not exist. So these considerations, while perhaps providing clues, do not really (seem to) tell much about other interactions.

What is the difference between interactions with massless objects, photons and "gravitons", and ones with mesons? In, say, Dirac's equation (schematically),

$$i\gamma_\mu \frac{d\psi}{dx_\mu} + m\psi + ie\gamma_\mu A_\mu \psi + g\pi\psi = 0, \qquad \text{(I.7.d.ii-1)}$$

if

$$\psi \Rightarrow \psi exp(ie\phi(x)), \qquad \text{(I.7.d.ii-2)}$$

a term $i\gamma_\mu \frac{d\phi}{dx_\mu}$ is added, but because A is a massless object, so with a solvable little group, the transformation also adds a term to it, canceling the first. But the pion, having a semisimple little group, does not get a term added. Thus we cannot argue that there should be a term giving an interaction with the pion to cancel a derivative-type term, as we can with the A.

The $\pi\psi$ term must have the same phase as ψ, the phase coming from gauge invariance (Poincaré invariance), and this requires that charge be conserved. The simplest way of achieving this would be for all charges to be integral multiples of a smallest one. While it might be possible for a particle to decay into a pair, neither of whose charges are integral multiples of the first, but which sum to its charge, each particle interacts with many different ones. Thus it would be difficult to have a complete system with charges that are not integral multiples of one.

Another problem with applying these arguments to require an interaction with the pion is that it is a scalar. If the transformation is for a space-dependent phase, the resultant field is a vector, the electromagnetic potential field. If the transformation also acts on the components of ψ, the field it leads to has three indices, and this is a gravitational connection. Arguments like these do not lead, at least directly, to scalar fields.

Thus greater insight and many more ideas are needed.

If there are some clues, there are even more questions raised by this discussion. Perhaps the questions are more valuable than the clues.

Chapter II

Moebius Groups

II.1 Conformal groups, geometry, and realizations

Since conformal transformations are so closely related to the geometry of our universe, thus to physics, it is useful, and for a full understanding of physics likely necessary, to study them and their implications in depth. This provides further insight into physics, geometry, group theory and other areas of mathematics, and suggests further lines of inquiry. It is also helpful to compare these for spaces of different dimensions and those with definite and with indefinite metrics, that is for which, as we see, the relevant curves are circles and equilateral hyperbolas, respectively, more generally (hyper)spheres and the hyperboloids that are their extensions to indefinite metric spaces.

Geometry (at least ours) can be (largely?) defined by this set of curves and surfaces and the transformations that generate it and under which it is invariant (sec. I.2.c, p. 17). Analyzing and visualizing this adds to our grasp of geometry, and of course of its curves and transformations.

These transformations and groups that they form can be realized in diverse ways. Some are considered elsewhere in this book; here we discuss them, and the curves, in greater depth for a special but important space, the plane, often the complex plane. These, and various combinations of them and of such types of extensions, might also lead to interesting insights and important new results; perhaps even a brief introduction can stimulate readers to see how they could occur and where they might lead.

Conformal transformations in the complex plane form a large set. In higher-dimensional spaces this set is more restricted (sec. III.1.a, p. 106). We often refer to the conformal group, especially when considered for general spaces, not merely two-dimensional, as the Moebius group [Burn (1991), p. 40; Neumann, Stoy and Thompson (1994), p. 19; Sattinger and Weaver (1986), p. 8; Schwerdtfeger (1979), p. 41], the name often used for the component connected to the identity. The full conformal group includes (disconnected) parts given by inversions, the Moebius group is the part that does not. Since we do not consider such inversions (which might be useful in providing understanding of physics) this is the relevant part.

Using, and comparing, different types of realizations and representations is usually enlightening. Representations of these groups (in higher-dimensional spaces) can be given using Clifford algebras, about which much is known, supplying useful tools. Vahlen matrices are representation matrices whose entries are members of Clifford algebras. We review them since we might benefit from available knowledge and techniques in the study of conformal representations, and again because different views of a subject can suggest.

While our main interest is Moebius groups for conformal groups of other spaces especially with dimension 3+1, the most easily visualized case is that of two dimensions, the complex plane. And the relevant curves are the most familiar. Studying how groups act on the plane and its curves illuminates larger groups. And many of our considerations are for a two-dimensional space (sec. I.3.a, p. 24); using the complex plane illustrates transformations that are fundamental in the approach. Also there is much known about this space. Of course the plane that we are interested in, the z, t plane, has one coordinate, t, imaginary, its metric is indefinite. That itself is enlightening. (There is a slight problem of notation. Since motion is generally denoted as being in the z direction we consider the z, t plane, where both z and t are real. But when we consider the complex plane, z is the complex coordinate. Thus $z = x + iy$, and for an indefinite metric space, $y \Rightarrow it$, so we consider motion in the x, t plane.)

Many properties of transversions, such as their action on spheres, hyperboloids, planes, and so on, are illustrated by their visualizable actions in a plane, on circles, hyperbolas, lines. So detailed study of these leads to greater intuition for other less easily visualized spaces. Such comprehension is useful, not only for geometry — and it leads to illumination of postulates underlying our geometry — but because actions of transversions differ from more familiar transformations, so help open minds, eliminate prejudices, and stimulate.

II.2 The conformal transformations in the complex plane

For a large part of the discussion it is necessary to consider only motion in one space dimension — in the two-dimensional x, t plane. But where possible arguments are presented so they can be extended to higher-dimensional spaces, specifically that of dimension 3+1.

Here the most interesting members of the conformal group are the special conformal transformations.

II.2.a Expressions for the conformal transformations

Conformal transformations are those that leave angles invariant, and the set of these (for each dimension) forms a group, the conformal group of that space (which includes inversions). In the complex plane it is well-known [Ford (1972), p. 1; Nehari (1975), p. 157; Neumann, Stoy and Thompson (1994), p. 214] that the most general such transformations (linear fractional transformations, bilinear transformation) are of the form (with all variables complex)

$$w = \frac{az + b}{cz + d}, \quad (ad - bc \neq 0), \tag{II.2.a-1}$$

the Moebius mapping [Lounesto (1997), p. 244].

As can easily be checked these include (for proper parameter values) rotations, translations, dilatations, plus others, special conformal transformations (transversions). These then are the transformations of the conformal group in the complex plane.

If $c = 0$ the transformation is merely a rescaling and translation, so leaving the form of the curve unchanged. For

$$ad - bc = 0, \tag{II.2.a-2}$$

which, with

$$z' = cz + d, \quad z = \frac{1}{c}(z' - d), \tag{II.2.a-3}$$

gives

$$w = \frac{az + b}{cz + d} = \frac{\frac{a}{c}(z' - d) + b}{z'} = \frac{a(z' - d + d)}{cz'} = \frac{a}{c}, \tag{II.2.a-4}$$

a circle of zero radius at $w = \frac{a}{c}$, rather then a transformation of the plane, the reason for requiring

$$ad - bc \neq 0. \tag{II.2.a-5}$$

Dividing all transformation parameters by d, defining

$$z' = \frac{c}{d}z, \quad b' = \frac{b}{d}, \tag{II.2.a-6}$$

we get

$$w = \frac{\frac{a}{c}z' + b'}{z' + 1} = b'\frac{\frac{a}{cb'}z' + 1}{z' + 1}, \tag{II.2.a-7}$$

$$w' = \frac{w}{b'} = \frac{a'z' + 1}{z' + 1}, \tag{II.2.a-8}$$

so in this form the transformation is a function of one complex parameter, a'.

As written these go from the z plane to the w plane. By letting $w = z'$ we indicate transformations of the z plane onto itself, providing relationships between its points. We can also consider it as a relabeling of points, a set of transformations between coordinate systems.

These transformations can also be given as (eq. III.1.c–3, p. 108)

$$x'_\mu = \sigma(x)^{-1}(x_\mu + c_\mu x^2), \tag{II.2.a-9}$$

$$\sigma(x) = 1 + 2c_\mu x_\mu + c^2 x^2; \tag{II.2.a-10}$$

their relationships are considered next.

II.2.a.i *Illustration of conformal transformations*

To illustrate such transformations we use

$$w = \frac{4z + 5 + 6i}{2z + 3}, \tag{II.2.a.i-1}$$

with real and imaginary parts,

$$wr = (8x^2 + 22x + 15 + 8y^2 + 12y)(x^2 + 12x + 9 + 4y^2)^{-1}, \tag{II.2.a.i-2}$$

$$wi = (y + 6x + 9)(x^2 + 12x + 9 + 4y^2)^{-1}, \tag{II.2.a.i-3}$$

and then we have for the real part of w

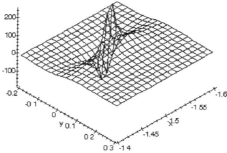

Figure II.2.a.i–1: Real part of the transformed coordinate,

and for the imaginary part

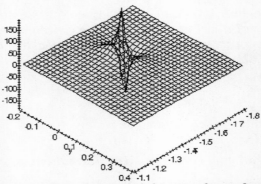

Figure II.2.a.i–2: Imaginary part of the transformed coordinate,

using
{w := (4*(x + I*y) + 5 + 6*I)/(2*(x + I*y) + 3); wr := simplify(evalc(Re(w)));
wi := simplify(evalc(Im(w))); plot3d(wi,x=-1.8..-1.1,y=-0.2..0.4,
color = black, style=line, axes=framed, grid=[30,30]); plot3d(wr, x = -1.9
.. -1.0,y=-0.2..0.3, color = black, style=line, axes=framed, grid=[20,20]);}.
     They have singularities as can be seen.

II.2.a.ii *Forms of conformal transformations and relationships*

Conformal transformations can be given in several ways (sec. III.1.b,
p. 107), and we summarize how these are related.
     The spaces that we are interested in are ones with indefinite met-
rics, since this includes our 3+1-space. But those with definite metrics,
Euclidean spaces, are more intuitive. We therefore review and express
many formulas for both types.
     While our emphasis is on transversions most of the discussion is
general, not limited to the subset of parameters giving only these. Also
these expressions refer to the complex plane, but by considering transformation parameters as members of a Clifford algebra they can be
generalized to other spaces (sec. II.5, p. 98).

II.2.a.iii *The conformal transformations in the complex plane*

Conformal transformations can be written using variables, z, that are
one-dimensional complex numbers. These are

$$w = \sigma(z)^{-1}(z + kz^2) = z\sigma(z)^{-1}(1 + kz), \qquad \text{(II.2.a.iii–1)}$$

$$\sigma(z) = 1 + 2kz + k^2 z^2 = (1 + kz)^2, \tag{II.2.a.iii-2}$$

or

$$w = (z^{-1} + k)^{-1} = z(1 + kz)^{-1}, \tag{II.2.a.iii-3}$$

where k is a complex number. By shifting the origin and changing the scale (a translation and a dilatation) it can be generalized to

$$w = (z^{-1} + k)^{-1} = z(1 + kz)^{-1} \Rightarrow (az + b)(cz + d)^{-1}. \tag{II.2.a.iii-4}$$

These latter two change the origin and scale, but not the resultant curve itself, which depends on one complex parameter. We shall have to see how forms of these curves, from specified types of initial curves, are dependent on this parameter.

How are these forms related in a general (n-dimensional) space? For these z and w are coordinate matrices, with inverses, and the coefficients in the transformations are themselves matrices (using Clifford algebras).

II.2.a.iv *Expressing a conformal transformation as a product*

A conformal transformation can be written (sec. III.1.d, p. 108) as a product of a translation, an inversion in a unit circle (and generalizations for other spaces) so leaving this curve invariant, a dilatation (for complex parameters this includes rotations in the complex plane), and a translation [Cartan (1995), p. 181]. These are

$$z_1 = z + \frac{d}{c}, \quad z_2 = \frac{1}{z_1}, \quad z_3 = kz_2, \quad w = z_3 + \frac{a}{c}, \tag{II.2.a.iv-1}$$

$$k = (bc - ad)c^{-2}. \tag{II.2.a.iv-2}$$

Translations given by these are of points in the complex plane, and while complex numbers can also be taken as vectors, the translated numbers do not give translated vectors of course. Then

$$w = \frac{a}{c} + k(z + \frac{d}{c})^{-1} = \{a(z + \frac{d}{c})c^{-1} + k\}(z + \frac{d}{c})^{-1}$$
$$= \frac{(acz + ad + c^2 k)}{c(cz + d)} = \frac{az + b}{cz + d}, \tag{II.2.a.iv-3}$$

so these forms for conformal transformations are equivalent.

II.2.a.v *Transformations in the complex plane formed by inversions*

Moebius transformation

$$w = \frac{az + b}{cz + d}, \quad (ad - bc \neq 0), \tag{II.2.a.v-1}$$

is a product of a translation, an inversion, and a translation (sec. III.1.d, p. 108), interchanging the inside and outside of a unit circle (or hyperbola), with the inversion, for the proper parameters, taking lines and circles into each other. We can also write ($c \neq 0$) [Churchill (1948), p. 58],

$$z' = \frac{1}{z}, \quad z'' = z' + c, \tag{II.2.a.v-2}$$

so

$$w = \frac{1}{z''} = (\frac{1}{z} + c)^{-1} = z(1 + cz)^{-1}, \tag{II.2.a.v-3}$$

writing the transformation as a product of an inversion, a translation, and an inversion, with the inversions thus about different points (else they would cancel). Translating gives w as a general conformal transformation, showing that it can be written as this product. For the complete conformal group we have to include translations, dilatations and rotations — they change the phase of z, but this is likewise done by a and c, these being complex, so rotations are included in this expression.

II.2.a.vi *Relating the two forms of the conformal transformations*

How are the two forms for the conformal transformation (eq. II.2.a.v-3; eq. III.1.c-3, p. 108) related? We use the expression for the inverse of a vector (eq. II.5.a-7, p. 99; sec. III.1.d, p. 108)

$$W_l = \frac{V_l}{V^2}, \quad W_l V_l = 1. \tag{II.2.a.vi-1}$$

Then, with multiplication being the scalar product,

$$\begin{aligned} w = (\frac{1}{z} + c)^{-1} &= (\frac{z}{z^2} + c)^{-1} = z^2(z + cz^2)^{-1} \\ &= z^2(z + cz^2)(z + cz^2)^{-2} \\ &= z^2(z + cz^2)(z^2 + 2czz^2 + c^2z^4)^{-1}, \end{aligned} \tag{II.2.a.vi-2}$$

so

$$w = (z + cz^2)(1 + 2cz + c^2z^2)^{-1}, \tag{II.2.a.vi-3}$$

which is the other form.

II.2.a.vii *Example of an inversion of an ellipse*

As an example of the effect of these transformations we consider an ellipse, then it inverted in the unit circle (drawn heavy), and the inverted ellipse translated (with the final one drawn dotted), to give the first two of these transformations {plot([[sin(t),cos(t),t = 0..2*Pi],[sin(t),2*cos(t),t = 0..2*Pi], [sin(t),0.5*cos(t), t = 0..2*Pi], [sin(t)+2.5,0.5*cos(t), t = 0..2*Pi]], style=[line,line,line,point],scaling=constrained,thickness=[3,1,1,1],color = black);},

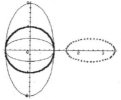

Figure II.2.a.vii–1: Ellipse, inverted then translated.

Next we take the inverted, translated ellipse and invert it again {dn := proc(t) local dn; dn := 2 + 5*sin(t) + 3*cos(t)**2 end; nm := proc(t) local nm; nm := sin(t) + 2.5 + 7.5*cos(t)**2 end; plot([[sin(t),cos(t),t=0..2*Pi],[nm(t)*((dn)**(-1)),2*((dn)**(-1)) *cos(t), t=0.. 2*Pi]], style=[line,point],scaling=constrained, thickness=[3,1,1,1], color =black);}. The unit circle is drawn again, but is so small it cannot be seen. So the curve resulting from this conformal transformation is

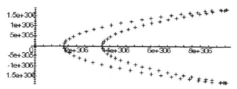

Figure II.2.a.vii–2: Curve given by a conformal transformation.

This transformation has a singularity (sec. I.6, p. 52) and its deformation of the ellipse is clear, as is its nonlinearity.

II.2.b   Invariance of angles

Conformal transformations are those that leave angles invariant (for the complex plane, relative phases). Clearly rotations, translations and dilatations (and inversions in planes and points) do not change angles. So we have to check the inversion in the unit circle (sphere, hypersphere, and similarly for indefinite metric spaces).

For vector V_i, in any of these spaces, inverse W_i resulting from this inversion is

$$W_i = \frac{V_i}{V^2}, \quad W_i V_i = 1. \tag{II.2.b.-1}$$

Vectors R and S, with inverses T and U, have angle given by

$$\sum R_i S_i = |R||S|cos\theta, \tag{II.2.b.-2}$$

therefore

$$|T| = \frac{1}{|R|}, \quad |U| = \frac{1}{|S|},$$

(II.2.b.-3)

$$\sum T_j U_j = \frac{\sum (R_j S_j)}{|R|^2 |S|^2} = \frac{|R||S|\cos\theta}{|R|^2 |S|^2}$$

$$= \frac{\cos\theta}{|R||S|} = |T||U|\cos\phi,$$

(II.2.b.-4)

so

$$\cos\phi = \cos\theta,$$

(II.2.b.-5)

and the angle is invariant. Thus it is invariant under all conformal transformations.

To check that these do leave phases fixed we use the sequence form (sec. II.2.a.iv, p. 71). The complex numbers give vectors whose ends they are, and these vectors are translated. Identical translations of two vectors do not change their relative phase. And it is clear that inverting the product of two complex numbers, or vectors, leaves the relative phase unchanged, as does multiplying them by the same complex number. Thus this sequence of operations does not change phases between vectors.

II.2.c   Why are there two special conformal transformations in the plane?

Transformation parameter c (eq. III.1.c-3, p. 108), or k (eq. II.2.a.iii-4, p. 71), is complex so there are two sets, given by c_r and c_i (plus of course combinations). What effect do these have, how do their actions differ? Why two? A straight line in the x, t plane is transformed into hyperbola (sec. II.3.d, p. 82)

$$x^2 - t^2 = \frac{1}{g^2},$$

(II.2.c.-1)

the path of an object with uniform acceleration (sec. I.3.b, p. 24). For the special case

$$t = 0, \quad x = \pm\frac{1}{g},$$

(II.2.c.-2)

which is the closest the hyperbola comes to the origin. There are two branches, and x moves toward ∞ (or $-\infty$) as $t \Rightarrow \pm\infty$.

Such hyperbolas have their vertices (turning points) on the axes, for this the x axis, and the closest approach to the origin is when $t = 0$. For hyperbola

$$t^2 - x^2 = \frac{1}{g^2},$$

(II.2.c.-3)

they are on the t axis. Here the smallest values of t are for $x = 0$. Of course this is not the path of a physical object, which would have to disappear for a time, as we see in {plot([sqrt(3 + x**2),-sqrt(3 + x**2)], x=-4..4, style=[line], scaling=constrained,thickness=[3],color=black);}

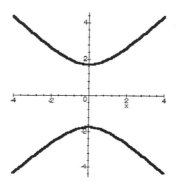

Figure II.2.c.-1: Hyperbola with axis parallel to t.

II.2.c.i *Freedom in choice of hyperbolas*

What freedom is there in picking such hyperbolas, and how are the choices related to values of the parameters? A hyperbola can be moved along the x axis, $x \Rightarrow x + s$, to give

$$x^2 + s^2 + 2sx - t^2 = \frac{1}{g^2}. \qquad \text{(II.2.c.i-1)}$$

We can also change the time of closest approach to the origin, $t \Rightarrow t + \tau$, so

$$x^2 - t^2 - \tau^2 - 2t\tau = \frac{1}{g^2}. \qquad \text{(II.2.c.i-2)}$$

Only the effect of transformations on hyperbolas with vertices on the axes need be considered since we can always shift them so this is true.

In addition axes can be rotated (by a boost), and g changed. These hyperbolas have different origins, different sizes (determined by g), and different orientations. This freedom requires two transversions — starting from an object at rest at an arbitrary point, we go to a moving frame, with arbitrary speed at time $t = 0$ (determining the orientation of the final hyperbola — its angle to x), and from that to arbitrary acceleration g. Thus two parameters are needed to specify a conformal transformation in a plane. How are their values related to the different hyperbolas?

II.2.c.ii *The meaning of the two special equilateral hyperbolas*

There are two special classes of equilateral hyperbolas which clarify the meaning of the parameter, one with axes parallel to the t axis, the other whose axes are parallel to x. Thus there are two sets of transversions (plus combinations). If both c's are nonzero, the axis of the resultant hyperbola is at an angle to the coordinate axes — however for physical transformations the range of the c's is limited (sec. I.6.b.ii, p. 54). As there are two sets of hyperbolas there must be two c's, and similarly for higher dimensions (with more axes to which those of hyperboloids can be parallel to).

What is the difference between these two classes? In the limit in which a hyperbola with axis along x becomes a straight line, it is the path of an object at rest (say at the origin). Thus arbitrary equilateral hyperbolas with axes parallel to x are paths of objects with constant acceleration that are tangent to the t axis at (say) $t = 0$, so instantaneously at rest at that time (although time translations are irrelevant to the meaning of the paths).

The other class has hyperbolic paths with axes along t, and in the limit these become straight lines perpendicular to t — parallel to x. These are paths, outside the light cone (here only two lines), so of objects of imaginary mass. We can also consider them as the lines along which those observers lie who start their clocks at the same time, but these are not paths of objects, nor are they our main interest here. Thus we need not consider transformations to such hyperbolas.

If a path for an object with real mass is rotated, so its axis is at an angle to x, the line tangent to it at (say) $t = 0$ is no longer along t, so no longer the path of an object at rest (in the same frame, although of course there is a frame for which it is). Thus the reason that — physically — there are two transversions is that, for a specified frame and origin they transform to hyperbolas that are differently oriented, so have different speeds at the origin (when transversions are regarded as giving accelerations). The c's then give the acceleration, and initial speed in the chosen frame. We can use Lorentz transformations to change frames, and set c's to zero (but not to cross the light cone for real parameters).

To check this we rotate the axes, so the c's, to give

$$c'_x = c_x cosh(\eta) + c_t sinh(\eta),\tag{II.2.c.ii–1}$$

$$c'_t = c_t cosh(\eta) + c_x sinh(\eta),\tag{II.2.c.ii–2}$$

and for $c'_x = 0$,

$$tanh(\eta) = -\frac{c_x}{c_t}, \quad tanh(-\eta) = \frac{c_x}{c_t},\tag{II.2.c.ii–3}$$

and similarly for $c'_t = 0$. With speed v

$$cosh(\eta) = \frac{1}{\sqrt{1-v^2}}, \quad sinh(\eta) = \frac{v}{\sqrt{1-v^2}}, \quad tanh(\eta) = v.\tag{II.2.c.ii–4}$$

So

$$v = -\frac{c_x}{c_t}. \qquad\qquad\qquad\qquad (\text{II.2.c.ii–5})$$

By boosting the object to this speed we transform to a system such that one of the parameters is zero, and the acceleration is determined by the other.

II.2.d   The effect of transversions on the axes

Transversions act on lines and their effect is determined by their parameters.

Taking $c_x = 0$ we see that the zero point of the x axis is unchanged, but the axis itself is transformed in a space-dependent way (sec. II.3.g.ii, p. 93). The latter is also true for the t axis, but its zero point is shifted to $t' = 0$, which is a function of x. Thus the curve $t = 0$ is no longer a straight line (observers at different points of space start their clocks at different times). The curve $x = 0$ remains a straight line ($x' = 0$ and $x = 0$ coincide for all values of t).

Conversely for $c_t = 0$, $t = 0$ remains a straight line (observers at all points of space start their clocks at the same time), while the curve $x' = 0$ is no longer straight (an observer at $x = 0$ finds its time intervals vary with time, although for this to have meaning it is necessary to measure time intervals in some invariant way).

In an inertial frame the curve $x = 0$ is the path of an object at rest at the origin. The curve $t = 0$ is the line on which objects lie whose clocks have the same initial time. Of course, the coordinate system can be rotated (boosted) so that these lines are different in different systems. After a special conformal transformation, with $c_x = 0$, objects that have the same starting time lie on an equilateral hyperbola. Taking them as accelerated, we notice that their clocks start at different times as seen by an inertial observer. The zero times seen by accelerated and inertial observers do not agree (except at one point). With $c_t = 0$, an object with $x' = 0$ has a position that varies with time as measured by an inertial observer. This is not surprising as it can be regarded as accelerated.

II.3   Transformations among lines and circles

Moebius transformations map straight lines into circles (which include straight lines as a special case) and circles (in definite metric spaces) into straight lines [Churchill (1948), p. 54; Nehari (1975), p. 156; Schwerdtfeger (1979), p. 11, 52; Yaglom (1968), p. 135], and similarly for non-definite metric spaces, clearly relevant to the interpretation of transversions as giving accelerations. Since this is a property of the transformations of geometry, thus of geometry, and clarifies the relationship of the conformal group to geometry, it is important to study and show it analytically. After the more familiar examples with circles, we consider indefinite-metric spaces so equilateral hyperbolas (sec. I.3.b, p. 24).

II.3.a   The equation for a circle

The general equation for a circle (with A, C, x, y, b_1, b_2 real) is [Ford (1972), p. 8]

$$A(x^2 + y^2) + b_1 x + b_2 y + C = 0, \qquad \text{(II.3.a-1)}$$

or

$$Azz^* + Bz + B^* z^* + C = 0, \qquad \text{(II.3.a-2)}$$

$$B = \frac{1}{2}(b_1 - ib_2), \qquad \text{(II.3.a-3)}$$

with

$$z = x + iy, \quad z^* = x - iy, \qquad \text{(II.3.a-4)}$$

$$x = \frac{1}{2}(z + z^*), \quad y = -\frac{i}{2}(z - z^*), \quad zz^* = x^2 + y^2. \qquad \text{(II.3.a-5)}$$

Its center, in the complex plane, is at

$$z_c = -\frac{B^*}{A}, \quad x_c = -\frac{b_1}{2A}, \quad y_c = -\frac{b_2}{2A}, \qquad \text{(II.3.a-6)}$$

(with B^* so all terms in the equation for the circle are real). The radius is

$$R = \sqrt{\frac{|B|^2 - AC}{A^2}}; \qquad \text{(II.3.a-7)}$$

for this to be real,

$$|B|^2 > AC, \qquad \text{(II.3.a-8)}$$

or if

$$B = 0, \quad \text{then} \quad AC < 0. \qquad \text{(II.3.a-9)}$$

If $A = 0$ (and $B \neq 0$), the curve is a straight line — straight lines are special cases of circles (and hyperbolas), planes are special cases of spheres, and so on for all dimensions. The center of the circle is at infinity if $A = 0$ (which is where it should be for a straight line — this then is the condition for straight lines), and then R is (of course) infinite.

Can coefficients be complex? Then the equation would be

$$A_r(x^2 + y^2) + b_{1r}x + b_{2r}y + C_r + i\{A_i(x^2 + y^2) + b_{1i}x + b_{2i}y + C_i\} = 0, \qquad \text{(II.3.a-10)}$$

which is the equation for two circles, with two sets of real parameters.

II.3.b   How transversions act on circles

From this circle, transformation

$$z' = (az + b)(cz + d)^{-1}, \tag{II.3.b-1}$$

$$z = (-dz' + b)(cz' - a)^{-1}, \quad z^* = (-d^*z'^* + b^*)(c^*z'^* - a^*)^{-1}, \tag{II.3.b-2}$$

gives circle,

$$\begin{aligned}
(Add^* &- Bc^*d - B^*cd^* + Ccc^*)z'z'^* \\
&+ (-Ab^*d + Ba^*d + B^*b^*c - Ca^*c)z' \\
&+ (-Abd^* + B^*ad^* + Bbc^* - Cac^*)z'^* \\
&+ Abb^* - Ba^*b - B^*ab^* + Caa^* = 0 \\
&= A'(x'^2 + y'^2) + b_1'x' + b_2'y' + C' = 0. \tag{II.3.b-3}
\end{aligned}$$

This is the equation of a circle since all coefficients (of x and y) are real. Conformal transformations then carry circles into circles. This can also be written (dropping primes)

$$\begin{aligned}
M(x^2 + y^2) + N(x + iy) + N^*(x - iy) + D &= 0 \\
= Mzz^* + Nz + N^*z^* + D &= 0 \\
= M(x^2 + y^2) + (N + N^*)x + i(N - N^*)y + D &= 0 \\
= M(x^2 + y^2) + 2N_rx - 2N_iy + D &= 0, \tag{II.3.b-4}
\end{aligned}$$

$$M = (Add^* - Bc^*d - B^*cd^* + Ccc^*), \tag{II.3.b-5}$$

$$N = -Ab^*d + Ba^*d + B^*b^*c - Ca^*c, \tag{II.3.b-6}$$

$$D = Abb^* - Ba^*b - B^*ab^* + Caa^*. \tag{II.3.b-7}$$

All coefficients, defined by this equation, are real, as are x and y.
For the transformed circle the center is at

$$z_c' = \frac{-N^*}{M} = (A^*bd^* - B^*ad^* - Bbc^* + C^*ac^*)M^{-1}, \tag{II.3.b-8}$$

and its radius is

$$R' = \sqrt{\frac{|N|^2 - MD}{M^2}}. \tag{II.3.b-9}$$

If $M = 0$ the transformed circle is a straight line, if $A = 0$ but not M, the transformed straight line is a circle. Thus for the proper parameters these transformations take lines to circles and circles to lines.

II.3.b.i *The effect of different parameter values*

As an example of how parameter values affect the curves we consider transformations of circles to circles using the form of the transformation (eq. II.2.a.iii-3, p. 71)

$$w = (z^{-1} + k)^{-1} = z(1 + kz)^{-1}, \qquad \text{(II.3.b.i-1)}$$

so

$$a = 1, \quad b = 0, \quad c = k, \quad d = 1, \qquad \text{(II.3.b.i-2)}$$

and it is k that we have to study.

For the transformed circle, with the untransformed one centered at the origin, and $A = 1$,

$$M = (A - Bk^* - B^*k + Ckk^*) = (1 + Ckk^*), \qquad \text{(II.3.b.i-3)}$$

$$N = B - Ck = -Ck, \qquad \text{(II.3.b.i-4)}$$

$$D = C. \qquad \text{(II.3.b.i-5)}$$

Its center and radius are

$$z_c = \frac{-N^*}{M} = \frac{Ck^*}{1 + Ckk^*}, \qquad \text{(II.3.b.i-6)}$$

$$R' = \sqrt{(|N|^2 - MD)M^{-2}} = \sqrt{[|Ck|^2 - (1 + Ckk^*)C](1 + Ckk^*)^{-2}}$$
$$= \sqrt{-C(1 + Ckk^*)^{-2}}, \qquad \text{(II.3.b.i-7)}$$

and C is always negative. Parameter values determine the radius and center of the transformed circle (in terms of those of the untransformed one). For fixed $|k|$ the phase determines the center. Two numbers are needed for a circle, its radius, given by $|k|$, and center for which we need also the ratio of real and imaginary parts, so two parameters fix a transformation.

II.3.b.ii *The transformations of a circle using rectangular coordinates*

We can also study the transformations using [Churchill (1948), p. 54]

$$z = x + iy, \quad w = u + iv = \frac{1}{z}, \qquad \text{(II.3.b.ii-1)}$$

so with

$$r^2 = x^2 + y^2, \qquad \text{(II.3.b.ii-2)}$$

$$q^2 = u^2 + v^2, \qquad \text{(II.3.b.ii-3)}$$

$$u = \frac{x}{r^2}, \quad v = -\frac{y}{r^2}, \qquad \text{(II.3.b.ii-4)}$$

$$x = \frac{u}{q^2}, \quad y = -\frac{v}{q^2}. \qquad \text{(II.3.b.ii-5)}$$

Now

$$Ar^2 + Bx + Cy + D = 0, \qquad \text{(II.3.b.ii-6)}$$

is a circle, for $A = 0$ a line. Likewise

$$Dq^2 + Bu - Cv + A = 0, \qquad \text{(II.3.b.ii-7)}$$

is a circle or line. However if $D = 0$ — the circle in the z plane passes through the origin — the curve in the w plane is a line, so this maps a circle to a line. For $A = 0$, this takes circles in the w plane into lines in the z plane. Thus

$$Dq^2 + Bu - Cv = \frac{D}{r^4}(x^2 + y^2) + B\frac{x}{r^2} + C\frac{y}{r^2} = 0, \qquad \text{(II.3.b.ii-8)}$$

so

$$D + Bx + Cy = 0, \qquad \text{(II.3.b.ii-9)}$$

a straight line. Considering the x and y axes, and u and v, symmetrically, $B = C$,

$$D'q^2 + u - v = 0, \qquad \text{(II.3.b.ii-10)}$$

and

$$D' + x + y = 0. \qquad \text{(II.3.b.ii-11)}$$

II.3.c Action of transformations using parametric forms for circles

Another way of studying these transformations is to write the equation for a circle using a parametric representation. For complex number z and real number v, which thus runs over a line (a, b, c, d are complex),

$$z = \frac{av + b}{cv + d}, \quad -\infty \le v \le \infty, \qquad \text{(II.3.c-1)}$$

is the equation of a circle — a straight line if $c = 0$. A map of a circle to a line is given by the inverse transformation,

$$v = \frac{b - dz}{cz - a}, \qquad \text{(II.3.c-2)}$$

so

$$z' = x' + iy' = \alpha v = \alpha\frac{b - dz}{cz - a}, \qquad \text{(II.3.c-3)}$$

goes from the circle in the z plane to the straight line for z'.

An example of this [Cartan (1995), p. 181] is

$$z = \frac{v - i}{v + i},$$
(II.3.c-4)

which maps the x (= v) axis to the unit circle. Points

$$v = 0, 1, \infty$$
(II.3.c-5)

are mapped to

$$z = -1, -i, 1.$$
(II.3.c-6)

Since this line is mapped to a line or circle, these three points show that it is a circle, and the unit circle.

Why do general conformal transformations map the real line into a circle? These leave the set of circles unchanged, with lines being a special case. Thus every line is mapped to a circle, or a line. By rotation of the axes any line can be taken as the real axis. Thus conformal transformations acting on it take it to a circle (perhaps one with infinite radius).

II.3.d   Transformations of hyperbolas

These give the transformations of circles (and straight lines). How do hyperbolas (including straight lines as a special case) transform under Moebius transformations? To establish notation we first review equilateral hyperbolas.

II.3.d.i   *Equilateral hyperbolas*

Hyperbolas that we are interested in (equilateral hyperbolas, paths for constant acceleration) have the general equation, with axis parallel to x (sec. II.2.c, p. 74),

$$\frac{(x - h)^2}{\alpha^2} - \frac{(t - k)^2}{\beta^2} = 1,$$
(II.3.d.i-1)

and for equilateral hyperbolas,

$$\alpha = \beta;$$
(II.3.d.i-2)

the vertices, the turning points, are at

$$x = h \pm \alpha, \quad t - k = 0.$$
(II.3.d.i-3)

The eccentricity is

$$\varepsilon = \frac{\sqrt{\alpha^2 + \beta^2}}{\alpha} > 1, \quad \text{so } \varepsilon = \sqrt{2},$$
(II.3.d.i-4)

if the hyperbola is equilateral.

The equation for an equilateral hyperbola can be written

$$(x^2 - t^2) - 2hx + 2kt + (h^2 - k^2 - 1) = 0. \qquad \text{(II.3.d.i-5)}$$

This can be obtained from the equation for a circle by setting $y \Rightarrow it$.

We choose the origin of the time axis so that $k = 0$ gives a particle moving to the left (for the other vertex, the right), reaching the closest point $(x_c = h + \alpha)$ to $x = 0$ at $t = 0$, going through zero velocity there and then reversing direction. This can always be achieved by taking coordinates belonging to a frame that is instantaneously at rest with respect to the object at $t = 0$, and choosing the x origin so the object is closest to it at that time. This point is at $(h \pm \alpha, 0)$.

The general equation (for $k = 0$) can be written, since we can rescale to get $\alpha = \beta$, and with g real,

$$x^2 - 2hx + h^2 - t^2 = \frac{1}{g^2}. \qquad \text{(II.3.d.i-6)}$$

We usually take $h = 0$.

A line needs only one parameter to give positions on it, so a hyperbola can also be given parametrically (eq. I.3.b.i-10, p. 26)

$$t = \frac{1}{g}sinh(g\tau), \quad x = \frac{1}{g}cosh(g\tau), \qquad \text{(II.3.d.i-7)}$$

where τ is the path parameter.

II.3.d.ii *Hyperbolas and rotations of the coordinate system*

Before studying the effect of conformal transformations it is useful to check the frame dependence. An inertial frame is taken into another such one by a Lorentz transformation,

$$x' = xcosh(\eta) + tsinh(\eta), \quad t' = tcosh(\eta) + xsinh(\eta). \quad \text{(II.3.d.ii-1)}$$

How do hyperbolas transform under these? They give

$$(x'^2 + h^2 - 2hx')\alpha^{-2} - (t'^2 + k^2 - 2kt')\beta^{-2} = 1 \Rightarrow$$
$$(x^2cosh^2(\eta) + t^2sinh^2(\eta) + h^2 - 2h(xcosh(\eta) + tsinh(\eta)))\alpha^{-2}$$
$$-(t^2cosh^2(\eta) + x^2sinh^2(\eta) + k^2 - 2k(tcosh(\eta) + xsinh(\eta)))\beta^{-2}$$
$$= x^2(\frac{cosh^2(\eta)}{\alpha^2} - \frac{sinh^2(\eta)}{\beta^2}) - t^2(\frac{cosh^2(\eta)}{\beta^2} - \frac{sinh^2(\eta)}{\alpha^2})$$
$$-2x(h\frac{cosh(\eta)}{\alpha^2} - k\frac{sinh(\eta)}{\beta^2}) + 2t(k\frac{cosh(\eta)}{\beta^2} - h\frac{sinh(\eta)}{\alpha^2})$$
$$+h^2 - k^2 = 1 - 2xt(\alpha^{-2} - \beta^{-2})cosh(\eta)sinh(\eta), \quad \text{(II.3.d.ii-2)}$$

which is a hyperbola with different size, shape and vertices.

For an equilateral hyperbola

$$(x' - h)^2 - (t' - k)^2 = \alpha^2, \qquad\qquad \text{(II.3.d.ii--3)}$$

so the transformed one is

$$x^2(cosh^2(\eta) - sinh^2(\eta)) - t^2(cosh^2(\eta) - sinh^2(\eta)) + h^2 - k^2$$
$$-2x(hcosh(\eta) - ksinh(\eta)) + 2t(kcosh(\eta) + hsinh(\eta))$$
$$= x^2 - t^2 - 2x(hcosh(\eta) - ksinh(\eta)) + 2t(kcosh(\eta) - hsinh(\eta))$$
$$\Rightarrow (x - (hcosh(\eta) - ksinh(\eta)))^2 - (t + (kcosh(\eta) - hsinh(\eta)))^2$$
$$= \alpha^2 - h^2 + k^2. \quad \text{(II.3.d.ii--4)}$$

In general it remains equilateral, but is translated. If the vertices are on the axis, and equidistance from t, $h = k = 0$, the hyperbola is not changed and is still equilateral. It is thus invariant under Lorentz transformations. This is equivalent to a circle centered at the origin, so invariant under rotations. The paths of uniformly accelerated objects are the same seen in all inertial coordinate systems. Thus our considerations do not depend on the particular inertial frame. Nor do they depend on the positions of the vertices, for we can always translate to get $h = k = 0$.

The general hyperbola is obtained by a rotation (boost), under which the distance for an indefinite metric space, $x^2 - t^2$, is (of course) invariant (as is the radius of a circle centered at the origin for rotations), with translations giving linear terms. Most properties can be found by using the simplest case, with vertices symmetrically placed on the axis, and that is all we study.

As an example of one branch of an equilateral hyperbola we have {plot([sqrt(-1 + (x+0.5)**2),-sqrt(-1 + (x+0.5)**2)], x=0.4..0.8,style=[line, line],thickness=[3,3],color=black);},

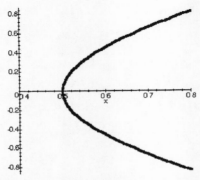

Figure II.3.d.ii-1: One branch of equilateral hyperbola.

II.3.d.iii *Obtaining hyperbolas from circles*

Another way of finding transformed hyperbolas is to start with circles centered at the origin

$$A(x^2 + y^2) + C = 0, \tag{II.3.d.iii-1}$$

$$Azz^* + C = 0. \tag{II.3.d.iii-2}$$

and in general

$$A((x - a)^2 + (y - b)^2) + C' = 0, \tag{II.3.d.iii-3}$$

or

$$A(x^2 + y^2) + b_1 x + b_2 y + C' = 0, \tag{II.3.d.iii-4}$$

where all coefficients are real.

Taking $y = it$ so

$$z = x + iy \Rightarrow x - t, \quad z^* = x - iy \Rightarrow x + t, \tag{II.3.d.iii-5}$$

$$x = \frac{1}{2}(z + z^*), \quad t = \frac{1}{2}(-z + z^*), \tag{II.3.d.iii-6}$$

the circle becomes a hyperbola, for real coordinates x and t,

$$A((x - a)^2 - (t - b')^2) + C' = 0, \tag{II.3.d.iii-7}$$

$$A(x^2 - t^2) + C = 0, \tag{II.3.d.iii-8}$$

$$A(zz^*) + C = 0. \tag{II.3.d.iii-9}$$

This last is the same equation as for a circle, but the expression for z in terms of x and t is different. If coefficients are taken imaginary, the equation becomes two, for two different hyperbolas. In general

$$A(x^2 - t^2) + b_1 x + ib_2 t + C = 0. \tag{II.3.d.iii-10}$$

Since all coefficients have to be real, $b_2 (= -ib_2')$ must be imaginary. So

$$B = \frac{1}{2}(b_1 - ib_2) \Rightarrow \frac{1}{2}(b_1 - b_2'). \tag{II.3.d.iii-11}$$

II.3.d.iv *The expression for conformal transformations in an indefinite-metric space*

This replacement of y by it can be used to write the conformal transformation in this hyperbolic space. With

$$z' = \sigma(z)^{-1}(z + cz^2), \tag{II.3.d.iv-1}$$

$$\sigma(z) = 1 + 2cz + c^2 z^2 = (1 + cz)^2, \tag{II.3.d.iv-2}$$

we can write

$$x' = \sigma(z)^{-1}(x + c_x(x + iy)^2), \qquad\qquad\text{(II.3.d.iv-3)}$$

$$y' = \sigma(z)^{-1}(y + c_y(x + iy)^2), \qquad\qquad\text{(II.3.d.iv-4)}$$

$$\sigma(z) = 1 + 2c_x x - 2c_y y + (c_x + ic_y)^2(x + iy)^2. \qquad\text{(II.3.d.iv-5)}$$

In an indefinite metric space $(y \Rightarrow it, c_y \Rightarrow ic_t)$,

$$x' = \sigma(z)^{-1}(x + c_x(x - t)^2), \qquad\qquad\text{(II.3.d.iv-6)}$$

$$t' = \sigma(z)^{-1}(t + c_t(x - t)^2), \qquad\qquad\text{(II.3.d.iv-7)}$$

$$\sigma(z) = 1 + 2c_x x + 2c_t t + (c_x - c_t)^2(x - t)^2. \qquad\text{(II.3.d.iv-8)}$$

The transformation here is a rotation, an orthogonal transformation. But we are taking one variable imaginary. However orthogonal groups can have complex parameters [Mirman (1995b), sec. 7.2, p. 124] so the resultant transformation is also a rotation, but a pseudo-rotation, a boost.

II.3.d.v *Transforming hyperbolas*

To find the transformed hyperbola we use a circle centered at the origin and consider the one resulting from the transformation (sec. II.3.b, p. 79) which becomes (with $y \Rightarrow it$, $N_i \Rightarrow -iN_i$) the hyperbola

$$M(x^2 - t^2) + 2N_x x + 2N_t t + D = 0. \qquad\text{(II.3.d.v-1)}$$

This is transformed by

$$z = \frac{-dz' + b}{cz' - a} = x - t = \frac{-d(x' - t') + b}{c(x' - t') - a}, \qquad\text{(II.3.d.v-2)}$$

$$z^* = x + t = \frac{-d(x' + t') + b}{c(x' + t') - a}, \qquad\text{(II.3.d.v-3)}$$

to

$$(Add^* + Ccc^*)z'z'^* + (-Ab^*d - Ca^*c)(x' - t')$$
$$+(-Abd^* - Cac^*)(x' + t') + Abb^* + Caa^* = 0, \quad\text{(II.3.d.v-4)}$$

$$(Add^* + Ccc^*)(x^2 - t^2) + x(-Ab^*d - Ca^*c - Abd^* - Cac^*)$$
$$+t(Ab^*d + Ca^*c - Abd^* - Cac^*) + Abb^* + Caa^* = 0. \quad\text{(II.3.d.v-5)}$$

The complex parameters of the transformation have to be interpreted as

$$c = c_r - c_i, \quad c^* = c_r + c_i, \qquad\qquad\text{(II.3.d.v-6)}$$

the sums of two real parameters, and similarly for the other parameters. Here

$$M = (Add^* + Ccc^*),$$
(II.3.d.v-7)

$$N = (-Ab^*d - Ca^*c),$$
(II.3.d.v-8)

$$N^* = (-Abd^* - Cac^*),$$
(II.3.d.v-9)

$$D = Abb^* + Caa^*.$$
(II.3.d.v-10)

For

$$a = 1, \quad b = 0, \quad c = k, \quad d = 1,$$
(II.3.d.v-11)

$$M = (A + Ckk^*),$$
(II.3.d.v-12)

$$N^* = -Ck^*, \quad N = -Ck,$$
(II.3.d.v-13)

$$D = C.$$
(II.3.d.v-14)

Thus the value of the complex parameter determines the size of the final hyperbola and also its position — the position of the vertices, which being on the axis is specified by a single number. So the parameter is given by two numbers and is complex.

What would result if the reality conditions did not hold? Then the transformed curve would be a sum of two equations of identical form, one with the parameters real, the other complex, so just giving two hyperbolas.

Now hyperbola

$$A(x^2 - t^2) - D = 0,$$
(II.3.d.v-15)

$$x^2 - t^2 = \frac{D}{A} = \frac{1}{g^2},$$
(II.3.d.v-16)

is given by a single real parameter. The vertices of this are on the axis, crossing the x axis at

$$x = \pm\frac{1}{g},$$
(II.3.d.v-17)

so it needs only a single parameter value. In general, a conformal transformation changes the value of g, the acceleration, and in addition moves the hyperbola.

II.3.d.vi *Invariance of null hyperbolas*

Does a Moebius transformation leave the light cone invariant? For these two lines

$$A(x^2 - t^2) + C = 0 = (x^2 - t^2),$$
(II.3.d.vi-1)

and transformed

$$M(x^2 - t^2) + N(x - t) + N^*(x + t) + D = 0,$$
(II.3.d.vi-2)

$$M = (A + Ckk^*) = A, \quad N = -Ck = 0,$$
(II.3.d.vi-3)

$$N^* = -Ck^* = 0, \quad D = C = 0.$$
(II.3.d.vi-4)

Thus the null hyperbola does remain invariant under conformal transformations, as others do not.

II.3.e How acceleration is changed by transversions

Transversions take paths of accelerated objects to other such paths. How does the value of the acceleration change under these transformations? An equilateral hyperbola, in the x, t plane (sec. II.3.d.iii, p. 85), is

$$x^2 - t^2 = -\frac{C}{A} = \frac{1}{g^2}, \qquad\qquad \text{(II.3.e-1)}$$

which is transformed to

$$M(x'^2 - t'^2) + 2N_r x' + 2N_i t' + D = 0, \qquad\qquad \text{(II.3.e-2)}$$

$$M = A(dd^* - \frac{1}{g^2}cc^*), \qquad\qquad \text{(II.3.e-3)}$$

$$N = A(-b^*d + \frac{1}{g^2}a^*c), \qquad\qquad \text{(II.3.e-4)}$$

$$N^* = A(-bd^* + \frac{1}{g^2}ac^*), \qquad\qquad \text{(II.3.e-5)}$$

$$D = A(bb^* - \frac{1}{g^2}aa^*), \qquad\qquad \text{(II.3.e-6)}$$

giving a hyperbola of different size (different acceleration) and translated (its vertices are shifted).

Shifting the origin so that we can compare hyperbolas,

$$x' = x'' + \eta, \quad t' = t'' + \tau, \qquad\qquad \text{(II.3.e-7)}$$

we have

$$(x''^2 - t''^2) + \eta^2 - \tau^2 + 2x''\eta - 2t''\tau$$
$$+2\frac{N_r}{M}x'' + 2\frac{N_i}{M}t'' + 2\frac{N_r}{M}\eta + 2\frac{N_i}{M}\tau + \frac{D}{M} = 0. \qquad \text{(II.3.e-8)}$$

So

$$2\eta + 2\frac{N_r}{M} = 0, \quad 2\tau - 2\frac{N_i}{M} = 0, \qquad\qquad \text{(II.3.e-9)}$$

which determines the shift of the origin. Then

$$(x''^2 - t''^2) + \eta^2 - \tau^2 + \frac{D}{M} = 0, \qquad\qquad \text{(II.3.e-10)}$$

giving

$$\frac{1}{g'^2} = \tau^2 - \eta^2 - \frac{D}{M} = \frac{N_i^2}{M^2} - \frac{N_r^2}{M^2} - \frac{D}{M}$$

$$= [\{(-b^*d + \frac{1}{g^2}a^*c) - (-bd^* + \frac{1}{g^2}ac^*)\}^2$$

$$-\{(-b^*d + \frac{1}{g^2}a^*c) + (-bd^* + \frac{1}{g^2}ac^*)\}^2](dd^* - \frac{1}{g^2}cc^*)^{-2}$$

$$-(bb^* - \frac{1}{g^2}aa^*)(dd^* - \frac{1}{g^2}cc^*)^{-1}, \qquad \text{(II.3.e-11)}$$

as the value for the acceleration, g', for the transformed hyperbola, in terms of that, g, for the untransformed hyperbola and the transformation parameters.

Taking

$$a = 1, \quad b = 0, \quad c = k, \quad d = 1,$$
(II.3.e-12)

this becomes

$$\frac{1}{g'^2} = \frac{1}{g^2} \frac{1}{1 - \frac{|k|^2}{g^2}}.$$
(II.3.e-13)

II.3.f When do straight lines go into straight lines?

The transformation from z to w (eq. II.2.a-1, p. 68), converts circles to circles, including taking straight lines into circles and conversely — lines are circles with infinite radii. Thus these transformations also take lines into lines (in indefinite-metric spaces, paths with constant velocity to ones also with constant velocity), and circles into circles (or in these spaces equilateral hyperbolas). For what parameter values does the type of curve remain the same?

A circle is a straight line if it passes through ∞. Thus if there is on the untransformed curve a point that is taken to $w = \infty$ the resultant curve is a straight line. This requires that the coefficient of zz^* (eq. II.3.b-4, p. 79),

$$M = Add^* - Bc^*d - B^*cd^* + Ccc^*,$$
(II.3.f-1)

be zero. In the equation of the original circle

$$Azz^* + Bz + B^*z^* + C = 0,$$
(II.3.f-2)

insert the coordinate for point

$$z = -\frac{d}{c}$$
(II.3.f-3)

to get

$$A(-\frac{d}{c})(-\frac{d^*}{c^*}) + B(-\frac{d}{c}) + B^*(-\frac{d^*}{c^*}) + C = 0.$$
(II.3.f-4)

Thus the transformation takes a circle into a line if for parameters d and c there is a point $z = -\frac{d}{c}$ on the circle — the circles passing through this point go into lines. Otherwise the resultant curve is a circle.

If the untransformed curve is a straight line, $A = 0$. Then for the transformed curve to also be one

$$B(-\frac{d}{c}) + B^*(-\frac{d^*}{c^*}) + C = 0,$$
(II.3.f-5)

giving the same condition on the parameters. In particular, since C is real, this requires, as it must,

$$C = Re[B(\frac{d}{c})]. \qquad \qquad (\text{II.3.f-6})$$

To get a hyperbola $y \Rightarrow it$, giving the same requirement, but using the equation for its curve (which is the same as that for a circle when expressed in terms of z, but this has a different relationship to x and t then the z for a definite-metric space has to x and y).

II.3.f.i   *Transformations leaving lines invariant form a subgroup*

There is a subset of the Moebius transformations that takes straight lines to straight lines. Does this form a subgroup? The transformation taking line I to line II has an inverse that takes line II to line I (because the Moebius transformations form a group) and this inverse is in the set. The product of the transformation that takes line I to line II with that taking line II to line III, takes line I to line III (because the Moebius transformations form a group). Further all three transformations belong to the set taking lines to lines (because the action of the product transformation can be found from the individual transformations, and since each takes a straight line to a straight line the product must). Thus this subset does form a subgroup.

II.3.f.ii   *Points at infinity that are mapped to infinity*

One important special case of a line going to a line is for the point at infinity to be mapped to that at infinity. The equation of a line is

$$[Ba^*d + B^*b^*c - Ca^*c]z' + [B^*ad^* + Bbc^* - Cac^*]z^{*\prime}$$
$$-Ba^*b - B^*ab^* + Caa^* = 0. \ (\text{II.3.f.ii-1})$$

Now (eq. II.3.b-2, p. 79) for $z' \Rightarrow \infty$,

$$z = -\frac{d}{c} \Rightarrow \infty; \qquad \qquad (\text{II.3.f.ii-2})$$

the point required to lie on the untransformed line is that at infinity. For this $c = 0$. Then

$$z = \frac{dz' - b}{a}, \qquad \qquad (\text{II.3.f.ii-3})$$

which is a change of scale plus a translation. Moebius transformations, except in such special cases, take lines to lines so take a finite point to that at infinity, and take the one at infinity to a finite point (sec. I.6, p. 52).

II.3.f.iii *Straight lines that are transformed to hyperbolas that are straight lines*

For a straight line going to a straight line rather than a hyperbola $M = 0$. Here (sec. II.3.b, p. 79), with all parameters real,

$$M = (Ad^2 - 2Bcd + Cc^2) = 0. \tag{II.3.f.iii-1}$$

Since the untransformed hyperbola (sec. II.3.d.iii, p. 85)

$$A(x^2 - t^2) + b_1 x - b_2' t + C = 0 \tag{II.3.f.iii-2}$$

is a straight line

$$A = 0, \quad b_1 x - b_2' t + C = 0, \tag{II.3.f.iii-3}$$

$$B = \frac{1}{2}(b_1 - b_2'), \tag{II.3.f.iii-4}$$

$$2Bd = Cc = (b_1 - b_2')d, \tag{II.3.f.iii-5}$$

which is satisfied by points

$$x = -t = -\frac{d}{2c}, \tag{II.3.f.iii-6}$$

$$-\frac{d}{2c}(b_1 + b_2') + C = 0. \tag{II.3.f.iii-7}$$

Thus transformations

$$z = \frac{-dz' + b}{cz' - a} = x - t = \frac{-d(x' - t') + b}{c(x' - t') - a}, \tag{II.3.f.iii-8}$$

with parameters d, c such that $-\frac{d}{2c}$ lies on the original straight line, take that line to another straight line (these also take a hyperbola to a straight line — the point must lie on the original curve). Other values take it to hyperbolas; for these the transformation from z to z' has a pole at z — the point z is that taken to infinity by the transformation, the requirement for the curve to be taken to a straight line.

II.3.f.iv *Where does infinity go under the transformation?*

If the transformation takes a line to a line, and a finite point on the former to infinity on the latter, where does it take the point at infinity of the original line? The transformation (eq. II.3.b-2, p. 79) of line

$$Bz' + B^* z'^* + C = 0, \tag{II.3.f.iv-1}$$

$$Re(Bz') + C = 0, \tag{II.3.f.iv-2}$$

$$B_r x' + B_i y' + B_i x' + B_r y' + C = 0, \tag{II.3.f.iv-3}$$

or, with the coefficients different, and all real,

$$B'_r x' - B'_i t' + B'_i x' - B'_r t' + C = 0, \qquad \text{(II.3.f.iv–4)}$$

takes point

$$z = -\frac{d}{c} \qquad \text{(II.3.f.iv–5)}$$

to z' which is at infinity (this depending on the transformation, not on the line). So

$$z' \Rightarrow \infty, \quad \text{for } z = -\frac{d}{c}, \qquad \text{(II.3.f.iv–6)}$$

and

$$z = \infty \text{ goes to } z' = \frac{a}{c}. \qquad \text{(II.3.f.iv–7)}$$

If all parameters are real and positive, $x' = +\infty$ goes to a negative point, as does $x' = -\infty$, this going also to $z = -\frac{d}{c}$ (however both of these cannot be on the same line), and the positive $x = \frac{a}{c}$ goes to $+\infty$. Since this is a condition on the real line, it is independent of y, so it is irrelevant whether the coordinate is y or t. Point

$$z' = \frac{b}{d} \text{ goes to } z = 0, \qquad \text{(II.3.f.iv–8)}$$

and

$$z' = 0 \text{ is taken to } z = -\frac{b}{a}. \qquad \text{(II.3.f.iv–9)}$$

Also

$$z = +\infty \text{ goes to positive } z' = \frac{a}{c}, \qquad \text{(II.3.f.iv–10)}$$

as does $z = -\infty$.

II.3.f.v *The transformation breaks the line into pieces*

What happens at

$$x' = \frac{a}{c}, \quad y' = 0, \qquad \text{(II.3.f.v–1)}$$

the point for which $z = \infty$? Now

$$z = \frac{-dz' + b}{cz' - a}; \qquad \text{(II.3.f.v–2)}$$

let

$$x' = \frac{a}{c} + \varepsilon, \qquad \text{(II.3.f.v–3)}$$

where ε changes sign at the point, so

$$y = 0, \quad x = (-d\frac{a}{c} + d\varepsilon + b)\varepsilon^{-1} = d + \frac{bc - da}{c\varepsilon}. \qquad \text{(II.3.f.v–4)}$$

$$x'' = \frac{(x-d)c}{bc-da} = \frac{1}{\varepsilon}. \tag{II.3.f.v-5}$$

Therefore x'', so x, goes to a large positive value, and thus to $+\infty$ (taking the denominator positive), then to $-\infty$ and then a large negative value. The same happens for x when it passes $-\infty$.

II.3.g How special conformal transformations differ

These transversion transformations, though they take lines to lines, are distinct from rotations and translations and combinations (and dilatations) because, for one reason, they interchange the point at infinity with a finite one. Thus they form a set of transformations different from and independent of the sets of rotations and the dilatation, and cannot be written as products of such transformations of the conformal group. The group really does have fifteen parameters.

II.3.g.i *Global transformations, and point-dependent ones*

Transformations of the similitude group — rotations, pure Lorentz transformations (boosts), translations and dilatations — are global, they are the same for every point of space. These are transformations between different coordinate systems, and the relations are the same for every point. However transversions (being nonlinear) vary over space (sec. I.5, p. 43). For ones considered here (complex parameters also give rotations), axes are stretched to take a finite point to (say plus) infinity, and points beyond are taken to the negative axis, which is compressed (the amount depending on position) so that points are properly mapped.

II.3.g.ii *The point-dependent dilatation of a line*

How do these space-dependent dilatations, the transversions, act on curves? These are (sec. III.1.c, p. 107)

$$z'_\mu = \sigma(z)^{-1}(z_\mu + c_\mu z^2), \tag{II.3.g.ii-1}$$

$$\sigma(z) = 1 + c_\mu z_\mu + c^2 z^2. \tag{II.3.g.ii-2}$$

Take first the light cone (for two dimensions a pair of intersecting lines),

$$z^2 = 0, \quad x = t. \tag{II.3.g.ii-3}$$

We then pick

$$c_x = -c_t, \quad c^2 = c_x^2 + c_t^2 = 2c_x^2, \tag{II.3.g.ii-4}$$

so

$$\sigma(z) = 1, \tag{II.3.g.ii-5}$$

and the transformation, for this particular cone, is the identity. For the line,

$$x = t + b; \tag{II.3.g.ii-6}$$

$$z^2 = x^2 - t^2 = 2bt + b^2. \tag{II.3.g.ii-7}$$

Then, with this same transformation,

$$\sigma(x) = 1 + c_\mu z_\mu + c^2 z^2 = 1 + c_x x + c_t t + c^2 z^2$$
$$= 1 + c_x b + c^2 (2bt + b^2), \tag{II.3.g.ii-8}$$

$$x' = \sigma(z)^{-1}(x + c_x(2bt + b^2)) = \sigma(z)^{-1}(t + b + c_x(2bt + b^2)), \tag{II.3.g.ii-9}$$

$$t' = \sigma(z)^{-1}(t - c_x(2bt + b^2)). \tag{II.3.g.ii-10}$$

Both coordinates are changed, and — since this is a transversion — in a space-dependent manner.

II.3.g.iii *Visualization of the action of transversions*

To visualize the effect of a transversion on a straight line consider the transformation from z' to z (eq. II.3.b-2, p. 79), and for a, b, c, d real and $y = 0$,

$$x = (-dx' + b)(cx' - a)^{-1}, \tag{II.3.g.iii-1}$$

for

$$x' = \frac{a}{c}, \tag{II.3.g.iii-2}$$

the transformed point is $x = \infty$. Transforming

$$x' = -\frac{a}{c} \quad \text{gives} \quad x = -\frac{da + bc}{2ac}. \tag{II.3.g.iii-3}$$

With

$$x' = \frac{a}{c} - \varepsilon, \tag{II.3.g.iii-4}$$

$$x = \frac{bc - da - dc\varepsilon}{-c\varepsilon}. \tag{II.3.g.iii-5}$$

So for small ε, $x = -\infty$ (say), x moves in as ε increases, and

$$x = 0 \quad \text{at} \quad x' = \frac{b}{d}. \tag{II.3.g.iii-6}$$

Thus the axis is stretched (here taken in the positive direction) starting from $\frac{b}{d}$ until $\frac{a}{c}$, for which $x = \infty$, and for larger values of x', with x starting at $-\infty$, the axis is compressed until x returns to the meeting point. The value obtained by the transformation from

$$x' = -\infty, \quad \text{is} \quad x = -\frac{d}{c}. \tag{II.3.g.iii-7}$$

So starting from this point, as x' moves in from $-\infty$, x moves in the positive direction. The x axis then can be considered in two parts. Beginning at the meeting point it is stretched so that a point on it goes to $x = \infty$, continuing gives $x = -\infty$, then x becomes less and less negative, and the axis is condensed. Point

$$x' = \infty, \text{ is taken to } \quad x = -\frac{d}{c}, \tag{II.3.g.iii-8}$$

the same value as for $x = -\infty$. Then as x' moves in from $-\infty$, x continues moving in the positive direction. Starting from $x' = -\frac{a}{c}$, and going in the negative direction, x also moves toward the negative direction, so that at the meeting point, the value of x obtained by moving from $-\infty$ in the positive direction, with that obtained from $x' = -\frac{a}{c}$ and going in the negative direction, are equal.

This behavior differs greatly from (space-independent) similitude transformations in breaking up the axis and distorting it.

The behavior in an indefinite metric space is the same, as we have taken y (or t) = 0, so the line is along the x axis and the metric is irrelevant. For general lines and transformation parameters the behavior would be similar, but there would be in addition rotations (including pure Lorentz transformations) and translations.

To illustrate the effect on the x axis we use

$$w = (x - 10)(5x - 10)^{-1}, \tag{II.3.g.iii-9}$$

and, plotting both the axis and the curve to which it is transformed we obtain the graph {plot([w, x = 0],x=-infinity..infinity, axes=framed, thickness=2,color=black,style = [point, line]);}

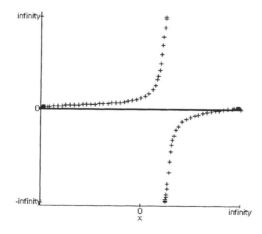

Figure II.3.g.iii–1: The x axis and a transformation of it.

II.3.g.iv *How the action of transversions and boosts on momentum differs*

Physically a transversion changes the magnitude of the four-momentum of (an accelerated) object and this change varies with time (so space); a boost changes the magnitude of the three-momentum with the change time-independent. This notes some of the differences between transversions and boosts.

We see then why the conformal group has fifteen sets of transformations, and fundamental differences between different types of transformations, and also consequences of transversions being nonlinear.

II.4 Elliptic coordinates

Since paths with constant acceleration are equilateral hyperbolas (which as we have seen are also closely related to our geometry) it is useful to have coordinates one of which is along these curves, the other along the curves perpendicular to the hyperbolas. These are elliptic coordinates (elliptic cylindrical coordinates) [Arfken (1970), p. 95; Margenau and Murphy (1955), p. 177] with the coordinate curves forming equilateral hyperbolas and ellipses — including the other dimensions, these types of (hyper-)cylinders. We just need coordinates in the z, t plane; those along x and y are rectangular coordinates.

The transformation between the two sets of coordinates is

$$x \Rightarrow x, \quad y \Rightarrow y, \tag{II.4-1}$$

$$t = \frac{1}{g} sinh(g\tau) sin(\eta), \tag{II.4-2}$$

$$z = \frac{1}{g} cosh(g\tau) cos(\eta), \tag{II.4-3}$$

$$0 \le g\tau, \quad 0 \le \eta < 2\pi. \tag{II.4-4}$$

The curves, in the z, t plane, with constant η are equilateral hyperbolas, those with constant τ are confocal ellipses.

These hyperbolas are the paths of objects with constant acceleration (sec. I.3.b, p. 24). Statefunctions of these objects are functions of only $g\tau$, giving plane waves (with phase constant along planes perpendicular to the hyperbolas).

II.4.a The invariants in elliptic coordinates

To obtain the invariant

$$\sum \frac{d^2}{dx_\mu^2} = \frac{d^2}{dt^2} - \frac{d^2}{dz^2} - \frac{d^2}{dx^2} - \frac{d^2}{dy^2} \tag{II.4.a-1}$$

in these coordinates we use

$$\frac{d}{dt} = \frac{d\tau}{dt}\frac{d}{d\tau} + \frac{d\eta}{dt}\frac{d}{d\eta}$$

$$= (cosh(g\tau)sin(\eta))^{-1}\frac{d}{d\tau} + g(sinh(g\tau)cos(\eta))^{-1}\frac{d}{d\eta}, \quad \text{(II.4.a-2)}$$

$$\frac{d}{dz} = \frac{d\tau}{dz}\frac{d}{d\tau} + \frac{d\eta}{dz}\frac{d}{d\eta}$$

$$= (sinh(g\tau)cos(\eta))^{-1}\frac{d}{d\tau} - g(cosh(g\tau)sin(\eta))^{-1}\frac{d}{d\eta}. \quad \text{(II.4.a-3)}$$

These have singularities. The reason is that the values of η giving these is for motion along an axis, so the slope is zero for one coordinate, infinite for the other. The corresponding value of τ is for a turning point, giving an infinite slope. In general η gives a scale factor, and can be chosen arbitrarily (unless the object is at rest, so its path is along t) thus the singularity it gives is irrelevant.

II.4.b   Dirac's equation in elliptic coordinates

The expression for the Dirac equation is obtained using these derivatives. The equation without interactions is

$$i\{y_o\frac{d\psi}{dt} - y_z\frac{d\psi}{dz} - y_x\frac{d\psi}{dx} - y_y\frac{d\psi}{dy}\} - m\psi = 0, \qquad \text{(II.4.b-1)}$$

and becomes

$$i\{[y_o(cosh(g\tau)sin(\eta))^{-1}\frac{d\psi}{d\tau} + g(sinh(g\tau)cos(\eta))^{-1}]\frac{d\psi}{d\eta}$$

$$-y_z[(sinh(g\tau)sin(\eta))^{-1}\frac{d\psi}{d\tau} - g(cosh(g\tau)sin(\eta))^{-1}]\frac{d\psi}{d\eta}$$

$$-y_x\frac{d\psi}{dx} - y_y\frac{d\psi}{dy}\} - m\psi = 0, \quad \text{(II.4.b-2)}$$

$$i\{y_o(cosh(g\tau)sin(\eta))^{-1} - y_z(sinh(g\tau)sin(\eta))^{-1}]\frac{d\psi}{d\tau}$$

$$+g[y_o(sinh(g\tau)cos(\eta))^{-1} + y_z(cosh(g\tau)sin(\eta))^{-1}]\frac{d\psi}{d\eta}$$

$$-y_x\frac{d\psi}{dx} - y_y\frac{d\psi}{dy}\} - m\psi = 0. \quad \text{(II.4.b-3)}$$

If we define y_τ and y_η as

$$y_\tau = y_o(cosh(g\tau)sin(\eta))^{-1} - y_z(sinh(g\tau)sin(\eta))^{-1}, \qquad \text{(II.4.b-4)}$$

$$y_\eta = g[y_o(sinh(g\tau)cos(\eta))^{-1} + y_z(cosh(g\tau)sin(\eta))^{-1}], \quad \text{(II.4.b-5)}$$

then

$$i\{y_\tau \frac{d\psi}{d\tau} + y_\eta \frac{d\psi}{d\eta} - y_x \frac{d\psi}{dx} - y_y \frac{d\psi}{dy}\} - m\psi = 0. \qquad \text{(II.4.b-6)}$$

But these y's are functions of the coordinates.

II.5   The Moebius group in terms of Clifford algebras

Moebius mappings [Lounesto (1997), p. 244; Porteous (1995), p. 251] are transformations of the complex plane of the form (sec. II.2.a.iii, p. 70)

$$w = \frac{az + b}{cz + d}, \quad (ad - bc \neq 0). \qquad \text{(II.5-1)}$$

They are conformal transformations of the entire complex plane including infinity. These conformal transformations of two-dimensional space (the Moebius transformations, giving the Moebius group [Beardon (1995), p. 20; Ketov (1997), p. 6]) can be extended to n dimensions. The matrices then have entries that are elements of a Clifford algebra [Ahlfors (1986); Jancewicz (1988), p. 30]. The space can also be a Minkowski space [Jancewicz (1988), p. 243], one with an indefinite metric. Thus [Elstrod, Grunewald and Mennicke (1987)] the conformal groups are closely related to, and can be given realizations in terms of, Clifford algebras [Cnops (1998); Lounesto (1997), p. 244; Lounesto and Latvamaa (1980)], and can be generalized using extensions of Clifford algebras [Ryan (1985)].

Such transformations, for the complex plane, are also known as linear fractional transformations, or homographic transformations [Churchill (1948), p. 57; Ford (1972), p. 1]. Clearly they are in the form of fractions, and are linear in the fractions, and z is linear in both the numerator and denominator. Homographic implies drawn or written in the same way, which is meaningful for the complex plane when transformed by a member of the conformal group.

II.5.a   The basis of a Clifford algebra

Clifford algebra, $\bar{C}_n$, is defined over an n-dimensional complex space, C^n, which when restricted to a real space, R^n, will be denoted by $\bar{C}_r^n$. This algebra has a basis [Boerner (1963), p. 267]

$$1, \hat{\underline{e}}_1, \ldots, \hat{\underline{e}}_n, \qquad \text{(II.5.a-1)}$$

where the elements $\underline{\hat{e}}_i$ obey the anticommutation relations

$$\{\underline{\hat{e}}_i, \underline{\hat{e}}_j\} = \underline{\hat{e}}_i\underline{\hat{e}}_j + \underline{\hat{e}}_j\underline{\hat{e}}_i = 2\delta_{ij}. \tag{II.5.a-2}$$

Also, for each element,

$$\underline{\hat{e}}_i^2 = 1. \tag{II.5.a-3}$$

Because of this every element of the algebra has an inverse. In 3+1-dimensional space the $\underline{\hat{e}}_i$ are the Dirac γ's, and in 3-space the Pauli σ's, for which these properties are well-known.

Now for every

$$\underline{z} = z_1\underline{\hat{e}}_1 + \ldots + z_n\underline{\hat{e}}_n \supset \bar{C}_n, \tag{II.5.a-4}$$

we have

$$\underline{z}^2 = -(z_1^2 + \ldots + z_n^2). \tag{II.5.a-5}$$

Not every nonzero vector in $\bar{C}_n$ has a multiplicative inverse in $\bar{C}_n$, for example

$$(\underline{\hat{e}}_1 + i\underline{\hat{e}}_2)^2 = 1 - 1 + \underline{\hat{e}}_1 \bullet \underline{\hat{e}}_2 + \underline{\hat{e}}_2 \bullet \underline{\hat{e}}_1 = 0. \tag{II.5.a-6}$$

However in the real subspace of this complex space, with coordinate vectors x, every vector is invertible, so

$$x^{-1} = \frac{x}{|x|^2}. \tag{II.5.a-7}$$

A basis for this space of the Clifford algebra over the reals, $\bar{C}_r^n$, consists of products of the $\underline{\hat{e}}_i$ vectors. We define the objects giving this basis as E_ν^p, a product of p $\underline{\hat{e}}_i'$s, so

$$E_\nu = E_\nu^p = \underline{\hat{e}}_{\nu_1} \ldots \underline{\hat{e}}_{\nu_p}, \tag{II.5.a-8}$$

for all sets of these indices, and all p's including 0 (for which the basis vector is 1), where

$$0 < \nu_1 < \nu_2 < \ldots \nu_p \leq n. \tag{II.5.a-9}$$

This ordering means that $p \leq n$.

II.5.b Clifford algebras and orthogonal-group Lie algebras

One reason Clifford algebras are important is that they give realizations of the Lie algebras of orthogonal groups [Boerner (1963), p. 269]. Thus the set of operators

$$\hat{\underline{\alpha}}_{ij} = \underline{\hat{e}}_i \bullet \underline{\hat{e}}_j \tag{II.5.b-1}$$

obeys (up to possible multiplicative constants) the commutation relations for the Lie algebra of orthogonal group SO(n).

Since the representations of Clifford algebras are known [Boerner (1963), p. 269] those of the orthogonal groups are determined — when these groups are realized as operators over their spaces of definition, which does not imply results for smaller spaces, as with the conformal group which is SO(4,2) realized over the 3+1-dimensional real space (sec. III.5.b, p. 162).

II.5.c The generalization of involutions

The involution (a transformation with unit square) for complex numbers is complex conjugation. Clifford algebras, over larger spaces, have three, generalizing this [Ahlfors (1986)]. These are defined by

$$E_\nu^{p*} = (-1)^p E_\nu^p, \tag{II.5.c-1}$$

$$E_\nu^{p-} = (-1)^{\frac{1}{2}p(p-1)} E_\nu^p, \tag{II.5.c-2}$$

$$E_\nu^{p+} = (-1)^{\frac{1}{2}p(p+1)} E_\nu^p. \tag{II.5.c-3}$$

For the complex plane, there is only a single $\hat{\underline{e}}_j$ and we can identify

$$\hat{\underline{e}}_j = i, \tag{II.5.c-4}$$

and any vector in this plane can be written in terms of the basis, 1 and i. Also $p = 1,2$, so

$$E_\nu^{1*} = (-1)^1 E_\nu^1 = -E_\nu^1, \tag{II.5.c-5}$$

$$E_\nu^{1-} = (-1)^{\frac{1}{2}1(1-1)} E_\nu^1 = E_\nu^1, \tag{II.5.c-6}$$

$$E_\nu^{1+} = (-1)^{\frac{1}{2}1(1+1)} E_\nu^1 = -E_\nu^1, \tag{II.5.c-7}$$

$$E_\nu^{2*} = (-1)^2 E_\nu^2 = E_\nu^2, \tag{II.5.c-8}$$

$$E_\nu^{2-} = (-1)^{\frac{1}{2}2(2-1)} E_\nu^2 = -E_\nu^2, \tag{II.5.c-9}$$

$$E_\nu^{2+} = (-1)^{\frac{1}{2}2(2+1)} E_\nu^2 = -E_\nu^2. \tag{II.5.c-10}$$

Thus, in the complex plane, two of these three involutions reduce to a single one, complex conjugation, the other merely the identity.

We can also state these in another way: E_ν^{p*} is obtained using transformation

$$\hat{\underline{e}}_i \Rightarrow -\hat{\underline{e}}_i, \tag{II.5.c-11}$$

and E_ν^{p-} by reversing the order of the terms in the E, with E_ν^{p+} the product of these two.

The inner product of two vectors is defined by

$$2(\underline{x}, \underline{y}) = \underline{x}\underline{y}^- + \underline{y}\underline{x}^-. \tag{II.5.c-12}$$

The square, and also any power and any polynomial of a vector is a vector. This holds also for fractional powers.

Writing the elements of the Clifford algebra as a, $b \supset \bar{C}$, and the scalar (inner) product of a and b as (ab), we have

$$(ab)^* = a^* b^*, \tag{II.5.c-13}$$

$$(ab)^+ = b^+ a^+, \tag{II.5.c-14}$$

$$(ab)^- = b^- a^-. \tag{II.5.c-15}$$

Thus $a \Rightarrow a^*$ is an isomorphism, the other two are anti-isomorphisms (whose meaning, given by reversing the order of the terms, is defined by these equations). For vectors

$$\underline{x}^+ = \underline{x}, \tag{II.5.c-16}$$

$$\underline{x}^* = \underline{x}^-. \tag{II.5.c-17}$$

This means that every nonzero vector is invertible and (eq. II.5.a-7, p. 99)

$$x^{-1} = \frac{x^*}{|x|^2}. \tag{II.5.c-18}$$

II.5.d The conformal group over this space

Conformal group transformations act on vectors, which include coordinates — these can be written in vector form as column vectors. The elements of the Clifford algebra form a complete set for these vectors. Thus a coordinate can be expressed as

$$x = x_i \hat{\underline{e}}_1, \tag{II.5.d-1}$$

where the coefficients x_i determine the particular coordinate.

We can now regard the transformations to act on these matrices x. Because the elements of Clifford algebras are invertible, each such coordinate matrix has an inverse.

II.5.d.i *The Moebius transformations*

The Moebius transformations are translations, $T_{\underline{w}}$

$$\underline{z} \Rightarrow \underline{z} + \underline{w}, \quad \underline{w} \supset C^n, \tag{II.5.d.i-1}$$

orthogonal transformations over the complex space, dilatations

$$\underline{z} \Rightarrow \lambda \underline{z}, \quad \text{with} \quad \lambda \supset C, \tag{II.5.d.i-2}$$

where C is the one-dimensional complex space (the complex plane) minus point 0, and the inversion (which is reflection in the unit sphere)

$$\underline{z} \Rightarrow \underline{z}^{-1}, \tag{II.5.d.i-3}$$

where these vectors include only those with inverses.

These transformations are realized by the matrices [Lounesto (1997), p. 248; Mack (1977), p. 5]

$$\begin{pmatrix} 1 & \underline{w} \\ 0 & 1 \end{pmatrix}, \quad \begin{pmatrix} a & 0 \\ 0 & a^{*-1} \end{pmatrix}, \quad \begin{pmatrix} \lambda^{\frac{1}{2}} & 0 \\ 0 & \lambda^{-\frac{1}{2}} \end{pmatrix}, \quad \begin{pmatrix} 1 & 0 \\ \underline{c} & 1 \end{pmatrix}, \tag{II.5.d.i-4}$$

where $\underline{w} \supset C^n$, λ giving the dilatation is real, $\underline{c} \supset R^{n,m}$, the space with metric (n,m), and $a \supset \text{Pin}(C^n)$, the covering group of $O(C^n)$, the orthogonal group over the n-dimensional complex space, including inversions, thus is a matrix. These matrices generate a group, the complex Vahlen group, $V(C^n)$.

II.5.d.ii   *The elements of the complex Vahlen group*

A general element of this group can be written as

$$\begin{pmatrix} a & b \\ c & d \end{pmatrix},$$ (II.5.d.ii-1)

where a, b, $\bar{c}$, $d \supset \bar{C}_n$, so are matrices. These matrix-valued elements of the Vahlen matrix belong to this Clifford algebra. However not all elements of this algebra are included, only those of the Clifford group (the set of Clifford numbers that can be written as products of nonzero vectors in R^n) plus 0, essentially the members of a Clifford algebra that have inverses [Porteous (1995), p. 140]. Also

$$ad^- - bc^- = 1,$$ (II.5.d.ii-2)

ab^- and cd^- both belong to R^n.

This gives the Moebius transformations as

$$\underline{x} \Rightarrow \frac{a\underline{x} + b}{c\underline{x} + d},$$ (II.5.d.ii-3)

where $\underline{x} \supset R^n$ (a vector in real space), and a, b, c, d belong to a Clifford algebra. This is a generalization of the 2-dimensional case for which a, b, c, d are numbers.

These transformations can transform a bounded region into an unbounded one (sec. I.6, p. 52).

II.5.e   The Lie Algebra of the Vahlen Group

The sets of these Vahlen matrices, those whose elements are members of a Clifford algebra, form representations of the conformal groups. These then give a realization of the Lie algebra of the group [Lounesto (1997), p. 252], as can be seen from their satisfaction of the commutation relations (sec. III.1.d.v, p. 112).

We then have

$$M_{\mu\nu} = \begin{pmatrix} -\frac{1}{2}\hat{\underline{e}}_{\mu\nu} & 0 \\ 0 & -\frac{1}{2}\hat{\underline{e}}_{\mu\nu} \end{pmatrix},$$ (II.5.e-1)

$$D = \begin{pmatrix} \frac{1}{2} & 0 \\ 0 & \frac{1}{2} \end{pmatrix},$$ (II.5.e-2)

$$P_\mu = \begin{pmatrix} 0 & \hat{\underline{e}}_\mu \\ 0 & 0 \end{pmatrix}, \qquad\qquad (\text{II.5.e-3})$$

$$K_\mu = \begin{pmatrix} 0 & 0 \\ \hat{\underline{e}}_\mu & 0 \end{pmatrix}. \qquad\qquad (\text{II.5.e-4})$$

These matrices are solvable (sec. I.7.c.iii, p. 61), as we expect, imply-ing that the transformations are inhomogeneous or gauge transforma-tions [Mirman (1995c), sec. 3.4, p. 43].

II.5.f The Dirac operator

Functions of a complex variable that are relevant to physics are ex-pected to be analytic, except perhaps for some singularities. But the space of physics is larger than the two-dimensional complex plane. We might expect that physical functions in 3+1-dimensional space are also analytic, except that this concept holds only for the complex plane. Thus how is it be generalized to larger dimensional spaces? There are different ways [Gilbert and Murray (1991), p. 87; Lounesto (1997), p. 255]. One, at least, has physical relevance, and is related to our present discussion. So we mention that.

II.5.f.i *The Cauchy-Riemann equations*

An analytic function,

$$f(z) = f(x, y) = u(x, y) + iv(x, y), \qquad\qquad (\text{II.5.f.i-1})$$

in the complex plane, obeys the Cauchy-Riemann equations,

$$\frac{du}{dx} = \frac{dv}{dy}, \quad \frac{du}{dy} = -\frac{dv}{dx}. \qquad\qquad (\text{II.5.f.i-2})$$

This can be written

$$d' f = 0, \qquad\qquad (\text{II.5.f.i-3})$$

where

$$d' = \frac{d}{dx} + i\frac{d}{dy}, \quad i^2 = -1. \qquad\qquad (\text{II.5.f.i-4})$$

Thus it is a sum of derivatives, each multiplied by an operator with square 1 or -1. This statement may seem trivial, but it does suggest a generalization [Brackx, Delanghe and Sommen (1982), p. 44].

For a function, regarded as a vector to two dimensions, to be analytic the requirement is that

$$div f = 0, \quad curl \ f = 0; \qquad\qquad (\text{II.5.f.i-5})$$

thus that the divergence and curl of the vector are both 0, is a restate-ment of the Cauchy-Riemann equations. Why should this be? If there is an electric charge at some point of the plane, the divergence of the electric field is nonzero at that point. Likewise a current-carrying wire produces a magnetic field, so the curl is not everywhere zero. In both cases there is a singularity, the fields are thus not everywhere analytic. In general for analyticity over the plane there can be no sources, not surprisingly.

II.5.f.ii *Generalization by means of the Dirac operator*

The Dirac operator in R^n is defined as

$$D = \sum_{j=1}^{n} \hat{\underline{e}}_j \frac{\partial}{\partial x_j}, \tag{II.5.f.ii–1}$$

with

$$\sum_{j=1}^{n} (\hat{\underline{e}}_j \frac{\partial}{\partial x_j})^2 = -1, \tag{II.5.f.ii–2}$$

$$(\hat{\underline{e}}_i \frac{\partial}{\partial x_i})(\hat{\underline{e}}_j \frac{\partial}{\partial x_j}) = -(\hat{\underline{e}}_j \frac{\partial}{\partial x_j})(\hat{\underline{e}}_i \frac{\partial}{\partial x_i}), \tag{II.5.f.ii–3}$$

and

$$D^2 \psi_n = 0, \tag{II.5.f.ii–4}$$

the Laplacian in R^n. These equations are satisfied by the Pauli σ's, and the Dirac y's. Hence the generalization of the Cauchy-Riemann equations to a space of (n, m) dimensions, with metric having n + signs and m - signs, is

$$Df = 0. \tag{II.5.f.ii–5}$$

The relevance of the Dirac operator here is that it gives this generalization.

These operators form a Clifford algebra, something well known for Dirac's equation. Thus Clifford algebras provide one generalization of analytic functions for spaces of arbitrary dimension. Of course, they are also used for representations of Moebius groups.

Dirac's equation (with no interactions, although putting these in might be interesting here)

$$D\psi = m\psi, \tag{II.5.f.ii–6}$$

gives one invariant of the generalized Poincaré group of this space, the Laplacian,

$$p^2 = m^2. \tag{II.5.f.ii–7}$$

And it determines the spin (for 3+1 space it is $\frac{1}{2}$), so its solutions thus also obey the equation for the second invariant [Mirman (1995b), sec. 6.2, p. 110].

This is but one example of the many relationships between conformal groups and function theory [Brackx, Delanghe and Sommen (1982); [Gilbert and Murray (1991); Ryan (1985)].

These emphasize then the value of the conformal group, often because it expands our views of group theory, and of many of the subjects that it touches, in mathematics, in physics, and elsewhere, thus stimulating further work and understanding.

Chapter III

Conformal Groups

III.1   The conformal group and physics

Conformal groups have been studied for many years [Barut, Budinich, Niederle and Raczka (1994); Campbell (1903), p. 32; Cobb and McCliment (1973); de Azcarraga and Izquierdo (1998), p. 354; Esteve and Sona (1964); Fillmore (1977); Ingraham (1998); Ketov (1997), p. 4; Macfadyen (1971a,b,1973a,b); Mack (1977); Schottenloher (1997), p. 12, 24; Yao (1967,1968,1971)], yet there is still much to learn. Although many applications to physics have been considered [Budinich and Raczka (1993); Carruthers (1971); Dirac (1936); Fulton, Rohrlich and Witten (1962a,b); Jaekel and Reynaud (1995); Kastrup (1966); Laue (1972); Piirainen (1996); Rosen (1968,1969); Weingarten (1973); Wess (1960)], it is quite likely that they have only been slightly exploited. Investigation of the groups could thus provide much insight into both physics and mathematics.

Because they differ from more familiar groups and have realizations and representations that are unlike ones to which we are accustomed, study of them expands our understanding of (not only) group theory, and suggests further lines of research in this, in other parts of mathematics (such as special functions), and in applications, including physics. In fact, group representation theory, as hinted in several places in this book, is much richer than usually realized. One purpose of these discussions is to point this out, certainly with the hope of stimulating study and application of this richness.

III.1.a   Conformal groups in general spaces

A conformal transformation is one that leaves angles invariant [Ingraham (1998); Porteous (1995), p. 245; Schottenloher (1997), p. 6]. In two dimensions, on the complex plane, the set of these is quite extensive;

however for larger dimensions, restrictions are greater. The set of these transformations in each dimension forms the conformal group listed next (here only for dimension 3+1). This is true for spaces of definite and also indefinite metric signatures [Haantjes (1937); Lounesto (1997), p. 244; Porteous (1995), p. 245].

We summarize the group, particularly its Lie algebra, and emphasize its less familiar, but likely quite interesting, properties and realizations especially those suggested by physics, perhaps for reasons beyond ones considered here.

Because it is different in many ways from more usual groups, and because of many novel aspects, especially for the representations that are relevant, there is a vast amount to be studied. Here we emphasize illustration of groups related to the conformal group and representations, to aid understanding, not only of them, but of more familiar cases by comparison with them, and to show (only) some areas that need, and likely can profitably, be studied. There is no attempt to provide a comprehensive treatment (the subject is too extensive for that, too much for even a start of such a treatment), but rather to indicate several of the realms that such treatments should explore.

III.1.b   The relevant forms of the transformations of the conformal group

To give representations of a group it is necessary to specify a decomposition into subgroups (usually little groups [Mirman (1999), sec. VI.2.c, p. 284]), upon whose representations those of the group are built. The ones considered here are unusual, for the conformal group is a homogeneous group, but we base its representations on those of the Poincaré subgroup [Mirman (1995a), sec. II.3.h, p. 45], an inhomogeneous group [Mirman (1995a), sec. XIII.4.b, p. 382], for it is this subgroup that is the transformation group of our geometry [Mirman (2001), sec. I.7.b, p. 37]. This leads to new ways of developing and using representations, which might by themselves be of more general use. Thus we outline the characteristics of these representations in some detail.

III.1.c   The transformations of the conformal group

The conformal group (limited to that of our 3+1-dimensional space, and its subspaces) has the transformations of the Poincaré group, whose Lie algebra generators are the homogeneous generators, forming the Lorentz algebra,

$$M_{\mu\nu} = -M_{\nu\mu}, \quad \mu, \nu = 1, 2, 3, 4 \qquad \text{(III.1.c-1)}$$

(... for higher-dimensional spaces), and momentum generators P_μ, the inhomogeneous ones. Generators M_{ij}, $i, j = 1,2,3$, are rotations, and

M_{0i} are the pure Lorentz transformations, the boosts. The dilatation is

$$x'_\mu = \rho x_\mu, \qquad\qquad (\text{III.1.c--2})$$

with generator D.

The word dilation comes from the Latin "dis" [Schwartzman (1994)] meaning "apart", or "away". It appears in many words, like "dissimilar", or "distance". The second part is from the Latin "latus", meaning "wide", this coming from the Indo-European root "stela-", "to extend". It thus implies an expansion. In this context it means (to widen apart so) an expansion, but by a factor that can be less than as well as greater than, or equal to, 1. Unless otherwise specified, as we do for transversions (sec. I.5, p. 43), a dilatation is an expansion by a constant, independent of position.

The other group elements are the special conformal transformations, transversions, with generators denoted by K_μ. The general conformal transformation is

$$x'_\mu = \sigma(x)^{-1}(x_\mu + c_\mu x^2), \qquad\qquad (\text{III.1.c--3})$$

$$\sigma(x) = 1 + 2c_\mu x_\mu + c^2 x^2. \qquad\qquad (\text{III.1.c--4})$$

The words transversal, transverse, come from the Latin "trans", this meaning "across" or "beyond" (as in transform, transitive, translate), this from the Indo-European root "tera-", to cross over, pass through [Schwartzman (1994)]. The second part comes from the Latin "versus", past participle of "vertere", to turn, or to go, from the Indo-European root "wer-", to turn, to bend. Transversion thus implies to go across, here to a different path or set of coordinates. Perhaps it becomes clearer by considering "version", so it implies to go across to a different version of the path, or coordinates. This suggests that the coordinates are not so much different, as a different version of the same ones. This is a concept to keep in mind as we study these transformations. Also transversions can be considered as point-dependent dilatations (sec. I.5, p. 43), so pictured as giving bending of axes — but with angles always maintained.

III.1.d   Transversions as inversions and translations

Transversions can be written, among other ways (sec. II.2.a.iv, p. 71), as [Ford (1972), p. 12; Ingraham (1998); Lyndon (1989), p. 144; Nehari (1975), p. 157; Neumann, Stoy and Thompson (1994), p. 214]

$$x'_\mu = T(-c)RT(c)x_\mu, \qquad\qquad (\text{III.1.d--1})$$

where R, which is not a member of a conformal group, is a mapping of the unit sphere (or unit circle in two dimensions, hyperspheres, and hyperboloids, for other spaces) interchanging the inside and out, but with the unit sphere itself unchanged,

$$Rx_\mu = \frac{x_\mu}{x^2},$$
(III.1.d-2)

and $T(c)$ is a translation,

$$T(c)x_\mu = (x_\mu + c_\mu).$$
(III.1.d-3)

This is then a translation, an inversion in the unit sphere, and a translation (sec. II.2.a.v, p. 71). Of course, this can be multiplied by other translations, by rotations and by dilatations. Thus we take the two translations as the same.

Transversions and inversions are then conjugate transformations [Mirman (1995a), sec. IV.3.a, p. 109], with the conjugation being the translation.

That angles remain the same under these transformations can be seen from the product of two unit vectors (sec. II.2.b, p. 73)

$$e_\mu f_\mu \Rightarrow \frac{e_\mu f_\mu}{e^2 f^2} = e_\mu f_\mu.$$
(III.1.d-4)

This invariance of angles is related to rotational invariance of space being unaffected by this inversion, as it must since the surface of reflection is a (hyper)sphere which cannot pick out a direction (the inversion takes the center of the sphere, and not some other point, to infinity).

III.1.d.i *How an inversion acts on a circle*

To illustrate the meaning of these transformations we consider the effect of an inversion on a circle (and the action on other conics, including hyperbolas, are similar). For the circle (sec. II.3.a, p. 78)

$$A(x^2 + y^2) + b_1 x + b_2 y + C = 0,$$
(III.1.d.i-1)

$$Azz^* + Bz + B^* z^* + C = 0,$$
(III.1.d.i-2)

and it becomes

$$A\frac{1}{zz^*} + B\frac{1}{z} + B^*\frac{1}{z^*} + C = 0,$$
(III.1.d.i-3)

$$A + Bz^* + B^* z + Czz^* = 0,$$
(III.1.d.i-4)

which gives

$$C(x^2 + y^2) + b_1 x + b_2 y + A = 0.$$
(III.1.d.i-5)

So a circle centered at the origin of radius

$$R^2 = -\frac{C}{A} \quad \text{goes to one at the origin with radius} \quad R^2 = -\frac{A}{C},$$

$$\text{(III.1.d.i–6)}$$

interchanging the inside and outside of the unit circle, leaving that invariant.

If the circle is displaced along the x axis a distance d

$$A((x-d)^2 + y^2) + C = 0 = A(x^2 + y^2) - 2Adx + Ad^2 + C, \quad \text{(III.1.d.i–7)}$$

the inverted circle is

$$(Ad^2 + C)(x^2 + y^2) - 2Adx + A = 0. \quad \text{(III.1.d.i–8)}$$

If

$$(Ad^2 + C) = 0, \quad \text{(III.1.d.i–9)}$$

the inverted circle is a straight line, showing again that the inversion can takes circles into lines. For the line

$$z + z^* = a(z - z^*) + d, \quad \text{(III.1.d.i–10)}$$

we obtain upon inversion

$$z + z^* = -a(z - z^*) + dzz^*, \quad \text{(III.1.d.i–11)}$$

which for $d \neq 0$ is a circle.

III.1.d.ii   *The inversion is not an element of a Lie group*

An inversion interchanging the inside and outside of a sphere is not an element of the conformal group, and in fact not an element of any Lie group. These groups have infinitesimal transformations, giving their algebras. But this inversion is global, there is no parameter on which it depends that can become infinitesimal (instead of the unit sphere one of a different radius can be picked, giving a parameter, but this is just a change of scale). Such an inversion gives a realization of the two-generator finite group, which is isomorphic to the groups of the reflection, or inversion through the origin, or time reversal [Mirman (1999), sec. VIII.2, p. 394]. However, unlike these, it does not leave magnitudes of intervals invariant, and the change of magnitude depends on position. Also these finite-group transformations can be regarded as elements of a Lie group by analytically continuing the transformations (thus a reflection can be obtained from the rotation group by taking angles complex, continuing the proper rotation group to the improper one [Mirman (1995b), sec. 7.2, p. 124; Streater and Wightman (1964), p. 13]). It still seems to be an open (but perhaps quite interesting) question whether this inversion of the sphere can be linked to a Lie group in some similar way.

Written in this form the transversions depend on a parameter, c_μ, and when it is zero, the transformation is the identity. Thus the Lie algebra operators are found from the first-order term in the expansion of the transformation in terms of this parameter.

III.1.d.iii   *These inversions do not change Poincaré invariance*

The space obtained by this transformation retains invariance under the Poincaré group. This invariance is not affected by translations, either before or after inversion, and the inversion leaves spheres (centered at the origin) and equilateral hyperboloids invariant under rotations and Lorentz transformations — there were no points distinguished on such surfaces before inversion, the inversion does not, so no points can be distinguished after. Thus transversions do not nullify Poincaré invariance (although the Poincaré transformations are, of course, different).

This is another reason for the relevance of the conformal group (sec. I.4, p. 36). The Poincaré group is equivalent to the geometry (at least locally). And the conformal group is the (largest) group that takes a Poincaré invariant space to another that is Poincaré invariant (it is the largest under which angles remain unchanged), so leaves the properties of the (local) geometry invariant.

III.1.d.iv   *Verification of the conformal group transformations*

The transformations of the conformal group are well-known, but how do we know that they are the correct ones? They are the transformations that leave angles invariant — similarities. What are these? Clearly translations, rotations, inversions, dilatations, and products, and these are all [Burn (1991), p. 191; Martin (1987), p. 196; Yale (1988), p. 71]. (Glides and screws are not included since space is symmorphic [Mirman (1999), sec. III.3, p. 137]).

Isometries form the Euclidean group in spaces with definite metric [Armstrong (1988), p. 136; Mirman (1999), sec. III.1, p. 132]; for our space with an indefinite metric the Poincaré group is the equivalent. An isometry is an "equal measure" that is a transformation that leaves lengths (and angles) invariant. The similitude group then is the set of isometries of the space, plus dilatations, which clearly are not isometries. With transversions — space-dependent dilatations (sec. I.5, p. 43) — we obtain the conformal group, and these also do not leave distances invariant.

Inversions in points (the origin) and planes (reflections) are familiar and obvious. However planes, and hyperplanes, are special cases of circles, spheres and hyperspheres (sec. I.2.c, p. 17). Thus inversions in these also leave angles invariant. Inversions in other curves or surfaces give distortions, so are not similarities.

Transversions are products of such inversions and of a translation. They are therefore members of the conformal group. Clearly products of these inversions and rotations around the origin (the center of the sphere used for the inversion) are included in the transformations, as are products with inversions in the origin, as well as products with dilatations. Further products with translations do not give additional members of the group.

Hence the set of conformal transformations are the translations, rotations, inversions (including reflections), dilatations, and transversions. These are the transformations that leave angles invariant, and the complete set that does.

III.1.d.v *The commutation relations of the algebra generators*

Conformal-algebra generators obey commutation relations

$$[D, P_\mu] = iP_\mu, \quad [D, M_{\mu\nu}] = 0, \qquad\qquad \text{(III.1.d.v–1)}$$

$$[D, K_\mu] = -iK_\mu, \quad [K_\mu, K_\nu] = 0, \quad [P_\mu, P_\nu] = 0, \qquad \text{(III.1.d.v–2)}$$

$$[K_\mu, P_\nu] = 2i(g_{\mu\nu}D - M_{\mu\nu}), \qquad\qquad \text{(III.1.d.v–3)}$$

where $g_{\mu\nu}$ is the metric,

$$[K_\lambda, M_{\mu\nu}] = i(g_{\lambda\mu}K_\nu - g_{\lambda\nu}K_\mu), \qquad\qquad \text{(III.1.d.v–4)}$$

$$[P_\lambda, M_{\mu\nu}] = i(g_{\lambda\mu}P_\nu - g_{\lambda\nu}P_\mu), \qquad\qquad \text{(III.1.d.v–5)}$$

and for the Lorentz subalgebra,

$$[M_{\kappa\lambda}, M_{\mu\nu}] = i(g_{\lambda\mu}M_{\kappa\nu} - g_{\kappa\mu}M_{\lambda\nu} - g_{\lambda\nu}M_{\kappa\mu} + g_{\kappa\nu}M_{\lambda\mu}) \quad \text{(III.1.d.v–6)}$$

[Carruthers (1971), p. 6; Schweber (1962), p. 45]. The algebra of the (inhomogeneous) conformal group is isomorphic to the (homogeneous) algebra so(4,2), which has the same number of generators and of commuting generators as su(2,2). This we show explicitly by giving the relationships between the operators for these algebras (sec. III.4.a.iii, p. 138).

Clearly the metric of such algebras cannot be positive-definite because our space on which the conformal group acts has metric +,-,-,-. Can the algebras be su(3,1) and so(5,1)? The metric has two positive signs, and four negative ones, because otherwise the signs in these commutation relations for the conformal generators would not all be correct. And these commutation relations are the ones obtained from the generators of the conformal transformations under which angles are invariant.

From these commutation relations we get (schematically)

$$exp(i\alpha D)P^2 exp(-i\alpha D) = exp(2\alpha)P^2, \qquad\qquad \text{(III.1.d.v–7)}$$

so mass is not invariant under the dilatation (the scale of all P's is changed by the same amount). The representation has a continuous mass spectrum (or the mass is zero). However mass is meaningless, only mass ratios have meaning (sec. I.2.a, p. 9). These ratios are invariant, thus this statement about the spectrum is physically meaningless.

III.1.d.vi   *Derivation of the Lie algebra generators for transversions*

The Lie algebra of the Poincaré group is well known. Thus we consider the generators for the transversions. The commutation relations follow from a realization (sec. III.1.e, p. 115).

Since the form of the transversions is unusual, being a fractional transformation, rather than an exponential, it is useful to review how an algebra is found from its group [Hamermesh (1962), p. 293]. Transformed coordinates, x_i', are functions of the original ones, x_j, and transformation parameters a_k,

$$x_i' = v_{ik}(x_j, a_k). \qquad \text{(III.1.d.vi-1)}$$

We write the parameters as real; for exponentials (which the eigenfunctions usually are) they are imaginary. Thus when the parameters are zero,

$$\frac{dx_i}{da_k} = u_{ik}(x_j), \quad dx_i = \sum u_{ik}(x_j) da_k; \qquad \text{(III.1.d.vi-2)}$$

functions u give the transformation. Consider now some arbitrary function $F(x)$, which changes under the transformation

$$dF = \sum \frac{dF(x)}{dx_i} dx_i = \sum u_{ik}(x_j) \frac{dF(x)}{dx_i} da_k. \qquad \text{(III.1.d.vi-3)}$$

Therefore

$$dF = \sum da_k X_k F, \qquad \text{(III.1.d.vi-4)}$$

and

$$X_k = \sum u_{ik} \frac{d}{dx_i}, \qquad \text{(III.1.d.vi-5)}$$

are the Lie algebra operators of the group.

For a translation

$$x_i' = x_i + c_i, \qquad \text{(III.1.d.vi-6)}$$

$$\frac{dx_i}{dc_k} = u_{ik} = \delta_{ik}. \qquad \text{(III.1.d.vi-7)}$$

Thus

$$X_k = p_l = \delta_{ik} \frac{d}{dx_i} = \frac{d}{dx_k}. \qquad \text{(III.1.d.vi-8)}$$

A rotation is

$$x' = x\cos\theta + y\sin\theta, \quad y' = y\cos\theta - x\sin\theta, \qquad \text{(III.1.d.vi-9)}$$

$$\frac{dx}{d\theta} = y, \quad \frac{dy}{d\theta} = -x, \quad u_{x\theta} = y, \quad u_{y\theta} = -x, \tag{III.1.d.vi-10}$$

so that

$$X = u_{x\theta}\frac{d}{dx} + u_{y\theta}\frac{d}{dy} = y\frac{d}{dx} - x\frac{d}{dy} = L_z. \tag{III.1.d.vi-11}$$

We now consider

$$z = \frac{aw + b}{cw + d}, \tag{III.1.d.vi-12}$$

for which the identity is given by parameter values (which are not all zero, since the finite transformation is fractional),

$$a = d = 1, \quad b = c = 0. \tag{III.1.d.vi-13}$$

Hence

$$\delta z = \frac{(1 + \delta a)w + \delta b}{\delta cw + (1 + \delta d)}. \tag{III.1.d.vi-14}$$

The algebra operators are found by setting these parameters to zero. So

$$X_a = w\frac{d}{dw}, \quad X_b = \frac{d}{dw}, \quad X_c = -w^2\frac{d}{dw}, \quad X_d = -w\frac{d}{dw}. \tag{III.1.d.vi-15}$$

Clearly X_b gives a translation, while X_a and X_d give rotations and dilatations, depending on the values of the parameters. Around the identity

$$z = \frac{(1 + \delta a)w}{1 + \delta d} = w(1 + \delta a - \delta d) + \dots . \tag{III.1.d.vi-16}$$

The parameter changes both the magnitude of w (a dilatation) and its phase (a rotation). The operator $X_c(= K)$ gives a transversion, and is nonlinear. Hence, with

$$w = x + iy, \tag{III.1.d.vi-17}$$

$$K_r = -(x^2 - y^2)\frac{d}{dx} - 2xy\frac{d}{dy}, \tag{III.1.d.vi-18}$$

$$K_i = (x^2 - y^2)\frac{d}{dy} - 2xy\frac{d}{dx}. \tag{III.1.d.vi-19}$$

The other expression for conformal transformations is

$$x'_\mu = \sigma(x)^{-1}(x_\mu + c_\mu x^2), \tag{III.1.d.vi-20}$$

$$\sigma(x) = 1 + 2c_\rho x_\rho + c^2 x^2. \tag{III.1.d.vi-21}$$

Thus, at the identity,

$$\frac{dx_1}{dc_1} = x^2 - 2x_1 x_1, \tag{III.1.d.vi-22}$$

$$\frac{dx_1}{dc_2} = -2x_1x_2. \qquad \text{(III.1.d.vi-23)}$$

In the limit with all transformation parameters zero,

$$\frac{dx_\mu}{dc_\nu} = u_{\mu\nu}(x_\rho) = \delta_{\mu\nu}x^2 - 2x_\mu x_\nu. \qquad \text{(III.1.d.vi-24)}$$

Then

$$X_\mu = K_\mu = \sum u_{\mu\nu}d_\nu = (\delta_{\mu\nu}x^2 - 2x_\mu x_\nu)d_\nu = x^2 d_\mu - 2x_\mu x_\nu d_\nu, \qquad \text{(III.1.d.vi-25)}$$

the transversion generators, in agreement with the complex form just found and with the ones given next except defined there with an i,

$$K_\mu = i(2x_\mu x_\nu d_\nu - x^2 d_\mu). \qquad \text{(III.1.d.vi-26)}$$

III.1.e Realization of conformal generators with internal parts

Differential expressions for generators have terms that act on internal variables, spin for the Lorentz group, but other generators can act on additional internal variables (whose interpretation is presently unclear). We want especially expressions for the K's, the generators giving accelerations. We use the little group [Mirman (1999), sec. VI.2.c, p. 284] that leaves the origin invariant, that is the group of internal transformations with operators δ, k_μ, and $\sigma_{\mu\nu}$, for the dilatations, special conformal transformations and Lorentz transformations; σ is the (generalization of the) spin operator. Operators of the algebra are then realized as [Mack and Salam (1969)]

$$P_\mu = id_\mu, \qquad \text{(III.1.e-1)}$$

$$M_{\mu\nu} = i(x_\mu d_\nu - x_\nu d_\mu) + i\sigma_{\mu\nu}, \qquad \text{(III.1.e-2)}$$

$$D = -ix_\nu d_\nu + \delta, \qquad \text{(III.1.e-3)}$$

$$K_\mu = i(2x_\mu x_\nu d_\nu - x^2 d_\mu + 2ix_\nu(g_{\mu\nu}\delta - \sigma_{\mu\nu}) + k_\mu); \qquad \text{(III.1.e-4)}$$

K_μ is nonlinear, as expected for an acceleration, and is what we use for the su(1,1) realization (sec. III.3.a.vii, p. 127). Also for the transversions there are terms that mix space and internal transformations.

III.1.f The number of commuting generators

Algebra so(4,2), to which that of the conformal algebra is isomorphic, is a complex extension of so(6), a rank-3 algebra. Yet so(4,2), and thus the conformal algebra, have four mutually commuting generators, for the conformal algebra called the momentum operators (sec. III.4.a.iv,

p. 139), and this is necessary for it to contain the Poincaré algebra as a subalgebra, and so to be a transformation algebra of 3+1 geometry. Also there is another set of four giving the transversions. How can a rank-3 algebra have more than three generators that commute?

While for a compact algebra the rank, the dimension of the Cartan subalgebra, gives the number of mutually commuting generators, this is not true for a noncompact algebra, as we see. The sign of the metric gives cancellation of terms that would otherwise add, so some commutators are zero but are not for the compact form. Boosts can be regarded as translations in momentum space, moving a point to a different one, that is one with a different momentum, as we see with the Lorentz algebra (sec. A.3.b, p. 230), for which the boost generator has translation eigenfunctions. Since momentum-space and position-space are duals, with operators differing only in the (signs and letters for the) realizations of the generators, not in their commutation relations, properties found for one hold for the other.

Thus the number of boosts — pseudo-rotations — gives the number of mutually commuting generators, that is momentum operators, which can be greater than that of the compact algebra. How many boosts are there? For the Lorentz group three, but these are related by rotations, thus there is one with the other two products of it and rotations, so these do not commute.

But for so(4,2) the so(4) compact subalgebra is of rank two, thus has two mutually commuting rotations, and from these we obtain two independent, so commuting boosts, which give momentum operators. A rotation, or boost, generator acts on two variables, so for a rotation in the xy plane, thus around z, it acts on these two (sec. A.3.a, p. 229). In three-dimensional space two such generators contain one coordinate in common, so cannot commute, which is one, among many, of the ways of seeing that they do not commute [Mirman (1995a), sec. X.4.b, p. 281]. But rotations in the xy and zw planes have different variables, so do commute.

Now for so(4,2) there are two "timelike" directions so two sets of boosts. Thus there are four mutually commuting operators, four momentum operators. But these momentum operators are neither rotations nor boosts, rather each is a sum of a rotation and a boost. Such sums can commute, while sums of rotations, as for so(6), cannot. But for su(2) $\Rightarrow$ su(1,1), the operators are not sums or differences, so do not allow cancellation for operators of the complex extension if there is none for the compact form.

This commutation is necessary for the group to have momentum generators, which are needed for it to be a transformation group of our geometry. This "accidental" equality of the required numbers, which is perhaps fortunate, must, and does, hold.

What is the effect of a generator being "timelike"? To obtain it we set $y \Rightarrow it$ and then the rotation generator

$$L_z = i(x\frac{d}{dy} - y\frac{d}{dx}) \Rightarrow B_z = -i(ix\frac{d}{dt} + it\frac{d}{dx}), \qquad \text{(III.1.f-1)}$$

changing the sign from negative to positive, so the operator from one for a rotation to that for a boost. For the three commutation relations of so(3) this changes the sign of one, converting the algebra from a compact to a noncompact one, from so(3) to so(2,1). The curves invariant under an so(3) operator are circles, and since a difference between two operators becomes a sum, these become equilateral hyperbolas for the complex extension (sec. II.3.d.i, p. 82), and in the basis states (spherical harmonics for so(3)), $sin(\theta) \Rightarrow sinh(\theta), cos(\theta) \Rightarrow cosh(\theta)$.

Since the algebras are isomorphic this also converts su(2) to su(1,1), but the action is different. The algebra operators are

$$E_z = \frac{1}{2}(E_1^1 - E_2^2) = \frac{1}{2}(z\frac{d}{dz} - z^*\frac{d}{dz^*}), \qquad \text{(III.1.f-2)}$$

$$E_+ = z\frac{d}{dz^*}, \qquad \text{(III.1.f-3)}$$

$$E_- = z^*\frac{d}{dz}, \qquad \text{(III.1.f-4)}$$

and this becomes the algebra of su(1,1) by multiplying the last two by i. The three compact operators count the difference in the number of z's and z^*'s in the polynomials that are the basis states and interchange these variables. But for the noncompact form the two step operators in addition shift the phase by $\frac{\pi}{2}$ for each basis state.

We see then that compact and noncompact algebras, even complex extensions of each other, can differ in ways as fundamental as the number of commuting generators. The conformal algebra although abstractly a complex extension of so(6) is quite different from it, and there is no complex extension of the conformal algebra that is isomorphic to so(6); although a complex extension perhaps could be written, it would not otherwise be related to the conformal algebra.

III.2 Labeling states and representations

To give representations of a group we have to provide labels for them and for the states, the state-labeling problem [Mirman (1995a), sec. I.6, p. 25]. There is freedom, for noncompact groups especially, and the nature of the representations depends on labels chosen (perhaps better, the type of labels is limited by the nature of the representations).

III.2.a   Using the Poincaré group to label states of the conformal group

The Poincaré group is the transformation group of space — although the entire conformal group is closely related to the geometry (sec. I.2, p. 9) — and the fundamental objects of the physics in that space are determined by its irreducible representations [Mirman (2001), sec. I.8.b, p. 48]. So here we consider representations whose states are basis states of the Poincaré subgroup — they are labeled by the eigenvalues of the invariants of the Poincaré group, and by its state labels. Thus, for the type of representations studied here, representations and states of the — homogeneous — conformal group are labeled by the eigenvalues of the invariants of the — inhomogeneous — Poincaré group. Then dilatation D and the transversions, given by the K's, which do not commute with the representation labels, mix Poincaré representations — but this does not imply that they mix Poincaré representations in the same way that boosts mix rotation representations (sec. I.4, p. 36). Such differences should be noted and understood for they might have important implications and value.

First we describe the representations and labels, then have to consider why it is possible to realize generators of a homogeneous group, here momenta, inhomogeneously (sec. III.3, p. 120).

III.2.b   The relationship between the Poincaré and conformal algebras

To find the Poincaré generators, which we here generically denote using $A_{\mu\nu}$, $\mu,\nu = 1,2,3,4$, in terms of the conformal ones, we use the metric

$$g_{\mu\nu} = (-1,-1,-1,1) = g_{\nu\mu}, \qquad\qquad \text{(III.2.b-1)}$$

and commutation relations

$$[A_{\mu\nu}, A_{\rho\sigma}] = i(g_{\rho\nu}A_{\mu\sigma} - g_{\mu\sigma}A_{\rho\nu}), \qquad\qquad \text{(III.2.b-2)}$$

giving the commutation relations, including those of the Poincaré group (sec. III.1.d.v, p. 112). The Lorentz generators are

$$M_{\mu\nu} = A_{\mu\nu} - A_{\nu\mu}. \qquad\qquad \text{(III.2.b-3)}$$

These M's are six of the conformal generators; the others are the six symmetric ones $A_{\mu\nu} + A_{\nu\mu}$, plus the three diagonal ones $A_{\mu\mu}$.

We note that

$$[A_{\mu\nu} - A_{\nu\mu}, A_{\mu\nu} + A_{\nu\mu}] \sim 2i(A_{\mu\mu} - A_{\nu\nu}), \qquad\qquad \text{(III.2.b-4)}$$

so, summing over μ,ν,

$$\sum [A_{\mu\nu} - A_{\nu\mu}, A_{\mu\nu} + A_{\nu\mu}] = 0. \qquad\qquad \text{(III.2.b-5)}$$

III.2.b.i *The Poincaré representation and state labels*

The (inhomogeneous) momentum operators are taken diagonal (giving the most familiar representations, although there are others [Mirman (1995c), chap. 2, p. 12]) so states have as labels their eigenvalues. Labels of the basis vectors of the homogeneous conformal group are thus provided by inhomogeneous generators. The first Poincaré invariant is $\sum P_\mu^2$, with eigenvalue

$$\sum p_\mu^2 = m^2, \tag{III.2.b.i-1}$$

a constant. To find the other Poincaré invariant [Schweber (1962), p. 45] we define the four-vector

$$w_\sigma = \frac{1}{2}\varepsilon_{\sigma\mu\nu\lambda}M^{\mu\nu}p^\lambda \tag{III.2.b.i-2}$$

where $\varepsilon_{\sigma\mu\nu\lambda}$ is the completely antisymmetric symbol [Mirman (1999), pb. IX.2.a-2, p. 471]. Then

$$W = -w_\mu w^\mu = \frac{1}{2}(M_{\mu\nu}M^{\mu\nu}p_\sigma p^\sigma - M_{\mu\sigma}M^{\nu\sigma}p_\nu p^\mu), \tag{III.2.b.i-3}$$

is the invariant. To interpret it we go to the frame (for a massive object, this is always possible) for which

$$p_i = 0, \quad i = 1, 2, 3, \tag{III.2.b.i-4}$$

then

$$W = \frac{1}{2}M_{\mu\nu}M^{\mu\nu}p_4 p^4 - M_{4\sigma}M^{4\sigma}p_4 p^4 = \frac{1}{2}m^2(M_{\mu\nu}M^{\mu\nu} - M_{4\sigma}M^{4\sigma})$$

$$= \frac{1}{2}m^2 M_{ij}M^{ij}. \tag{III.2.b.i-5}$$

In this frame

$$w = m(0, M_{23}, M_{31}, M_{12}), \tag{III.2.b.i-6}$$

where these M's are the generators of the rotation group. Thus the eigenvalue is

$$W_l = -w_\mu w^\mu = m^2 l(l + 1); \tag{III.2.b.i-7}$$

l is the angular momentum quantum number. Hence the values of the invariants are the mass (squared) given by the set of eigenvalues of the P's (with no implication here that this is an actual mass), and the total angular momentum (times the mass squared) in the rest frame. This does not imply that there is a physical angular momentum — here it is just a state label — or that there is a physical rest frame beyond its present meaning of a particular set of values of p_i.

III.2.b.ii *The conformal representation-labeling invariants*

The first step in finding representations is to determine the Casimir invariants and their eigenvalues. Since the Lie algebra of the conformal group is isomorphic to that of SO(4,2) whose Cartan subalgebra has the same dimension as that of SU(2,2), there are 3 invariants. But while the invariants of those algebras are functions of operators on 6 real variables, and 4 complex ones, respectively, for the conformal group they are functions of operators on only 4 real variables, the four coordinates of Minkowski space. And since the group is noncompact, there are other, nonpolynomial ones, as with the Lorentz and Poincaré groups, which for these distinguish the forward and backward light cones and their interiors. Also the operators are realized very differently from (the familiar) ones of those algebras, some being inhomogeneous, others nonlinear, rather than all homogeneous, and all linear, as for the usual semisimple realizations. The invariants require detailed investigations, and the results are likely to be interesting.

III.3 Groups with properties similar to those of conformal groups

The conformal group has unusual properties and is also somewhat complicated. It is therefore useful to study other groups that are similar but simpler, and that help clarify why the conformal group has, can have, need have, such properties. The simplest semisimple algebra is su(2), quite familiar, but still capable of being cast into a form similar to the conformal group. While study of this helps elucidate properties of the conformal group it also emphasizes that even the most elementary, familiar systems may allow highly different configurations if looked at with some originality.

The conformal group has fifteen parameters, but is not the only one. Relating it to the others is not only useful, but probably necessary. These we also consider.

III.3.a The su(1,1) algebra of D, P and K

To study conformal group representations we start with the simplest case (which is noncompact) for which we can form representations with somewhat equivalent properties. (An analogous treatment can be given for su(2).) The three operators of an su(1,1) algebra have commutation relations,

$$[D,P] = iP, \qquad\qquad\qquad\text{(III.3.a-1)}$$

$$[D,K] = -iK, \quad [K,P] = 2iD. \qquad\qquad\text{(III.3.a-2)}$$

What realizations of it are similar to those of the conformal group?

We realize the operators, for one dimension, as

$$P = i\frac{d}{d\tau},$$
(III.3.a-3)

$$K = i\tau^2\frac{d}{d\tau},$$
(III.3.a-4)

$$D = -i\tau\frac{d}{d\tau}.$$
(III.3.a-5)

Here P is realized, not homogeneously as is usually done, but inhomogeneously. This is possible because K is realized nonlinearly, with a τ^2, rather than a τ (sec. III.5.b, p. 162). There is a required connection between inhomogeneous realizations of some operators and nonlinear realizations of others. It is this that makes these realizations interesting.

There is also the dual to this realization

$$P = i\tau,$$
(III.3.a-6)

$$K = i\tau\frac{d^2}{d\tau^2},$$
(III.3.a-7)

$$D = i\tau\frac{d}{d\tau},$$
(III.3.a-8)

and these have the action on an arbitrary function of τ,

$$KPf(\tau)-PKf(\tau) = -\tau^2 f'' - 2\tau f' + \tau^2 f'' = -2\tau f' = 2iDf. \quad \text{(III.3.a-9)}$$

If all operators are realized homogeneously, the realization and its dual are the same. Here they are different.

III.3.a.i *Types of representations*

There are three sets of representations, for each of the two realizations, that with D diagonal, with P diagonal and with K diagonal. Again for homogeneous realization of all operators these are the same — they differ only in labeling. But here operators are realized differently leading to three representation types.

III.3.a.ii *Representations for which D is diagonal*

If D is diagonal the structure of the representations for the realization and its dual are the same (as we expect). The states have the form

$$|n) = N_n\tau^n,$$
(III.3.a.ii-1)

given by the integers n; N_n is a normalization factor. This requires a definition of normalization, thus a range of τ and a measure, which could add further richness to the types of representations; we do not study this here, but note the question for it might be worth study.

The action of the operators is then

$$D|n) = -i\tau\frac{d|n)}{d\tau} = -in|n). \tag{III.3.a.ii-2}$$

$$P|n) = i\frac{d|n)}{d\tau} = in\frac{N_n}{N_{n-1}}|n-1), \tag{III.3.a.ii-3}$$

$$K|n) = i\tau^2\frac{d|n)}{d\tau} = inN_n\tau^{n+1} = in\frac{N_n}{N_{n+1}}|n+1). \tag{III.3.a.ii-4}$$

Since the states differ by integers a fractional power of τ could be put into the normalization factor, which would then be space-dependent.

For the dual realization,

$$D|n) = i\tau\frac{d|n)}{d\tau} = in|n), \tag{III.3.a.ii-5}$$

$$P|n) = i\tau|n) = i\frac{N_n}{N_{n+1}}|n+1), \tag{III.3.a.ii-6}$$

$$K|n) = i\tau\frac{d^2|n)}{d\tau^2} = in(n-1)\frac{N_n}{N_{n-1}}|n-1). \tag{III.3.a.ii-7}$$

While the notation for normalization factors is the same for both types, the factors may not be, as the action of the operators is different. This is something that has to be looked at.

To check these note that

$$[D,P]|n) = iP|n) = -n\frac{N_n}{N_{n-1}}|n+1)$$

$$= n(n-1)\frac{N_n}{N_{n-1}}|n-1) - n^2\frac{N_n}{N_{n-1}}|n-1), \tag{III.3.a.ii-8}$$

$$[D,K]|n) = -iK|n) = n\frac{N_n}{N_{n+1}}|n+1)$$

$$= (n(n+1) - n^2)\frac{N_n}{N_{n+1}}|n+1), \tag{III.3.a.ii-9}$$

$$[K,P]|n) = 2iD|n) = -2n|n)$$

$$= \frac{N_n}{N_{n-1}}n(n-1)\frac{N_{n-1}}{N_n}|n) - n(n+1)\frac{N_n}{N_{n+1}}\frac{N_{n+1}}{N_n}|n), \tag{III.3.a.ii-10}$$

which are independent of normalization.

What conditions are there on these? The Casimir operator in terms of the generators,

$$C = D^2 - \frac{1}{2}(PK + KP), \tag{III.3.a.ii-11}$$

is realization independent (it is determined solely by the commutation relations). If the representation is unitary, the step operators are hermitian conjugates. Then every term is positive definite. In both of these realization forms there is an operator increasing the eigenvalue of D, the other decreasing it. But unlike usual realizations, these are not hermitian conjugates; D is antihermitian, while P and K are not conjugate. Thus we cannot conclude that there must be a state for which each operator gives zero — that there are largest, and smallest (most negative), values of n.

For the realization for which P is a differential operator

$$C|n) = D^2|n) - \frac{1}{2}(PK + KP)|n)$$

$$= -n^2|n) + \frac{1}{2}n(n+1)\frac{N_{n+1}}{N_n}\frac{N_n}{N_{n+1}}|n) + \frac{1}{2}n(n-1)\frac{N_{n-1}}{N_n}\frac{N_n}{N_{n-1}}|n)$$

$$= -n^2|n) + \frac{1}{2}\{n(n+1) + n(n-1)\}|n) = 0. \qquad \text{(III.3.a.ii-12)}$$

Thus the Casimir invariant imposes no conditions. For the dual realization to this again it imposes no conditions,

$$C|n) = D^2|n) - \frac{1}{2}(PK + KP)|n)$$

$$= -n^2|n) + \frac{1}{2}n(n-1) + n(n+1)|n) = 0. \qquad \text{(III.3.a.ii-13)}$$

That the eigenvalue of the Casimir operator is zero is not surprising since there is no value for it to be a function of. It would be a function of the boundary value of n, but here there is none.

For usual realizations of su(2) and su(1,1) there are an infinite number of representations, whose states are mixed among themselves by the group operators, but not with states of other representations; here all representations have the same form, so are essentially identical. The only possibility of differences is in the normalization. Of course, this algebra is very simple.

It is interesting that the states form a discrete set, so the expectation values of D also are discrete (appendix B, p. 246).

III.3.a.iii *Representations with P diagonal*

The next type of representation is for P diagonal. Then

$$P|p) = i\frac{d|p)}{d\tau} = -pexp(ip\tau). \qquad \text{(III.3.a.iii-1)}$$

This gives (with any normalization suppressed)

$$K|p) = i\tau^2\frac{d|p)}{d\tau} = -\tau^2 pexp(ip\tau), \qquad \text{(III.3.a.iii-2)}$$

$$D|p) = -i\tau\frac{d|p)}{d\tau} = \tau p \, exp(ip\tau).$$ (III.3.a.iii-3)

Also

$$PK|p) = -\frac{d}{d\tau}(\tau^2\frac{d|p)}{d\tau}) = \tau^2 p^2 exp(ip\tau) - 2i\tau p \, exp(ip\tau),$$
(III.3.a.iii-4)

$$KP|p) = -\tau^2\frac{d}{d\tau}\frac{d|p)}{d\tau} = \tau^2 p^2 exp(ip\tau),$$ (III.3.a.iii-5)

$$D^2|p) = -\tau\frac{d}{d\tau}\tau\frac{d|p)}{d\tau} = (\tau^2 p^2 - i\tau p)exp(ip\tau).$$ (III.3.a.iii-6)

Its Casimir invariant

$$C|p) = D^2|p) - \frac{1}{2}(PK + KP)|p)$$

$$= (\tau^2 p^2 - i\tau p - \frac{1}{2}(2\tau^2 p^2 - 2i\tau p))|p) = 0,$$ (III.3.a.iii-7)

gives no conditions.
    To check

$$[D,P]|p) = iP|p) = -\tau p^2|P) + \tau p^2|P) - ip|p) = -ip|p),$$ (III.3.a.iii-8)

$$[D,K]|p) = -iK|p) = (-\tau^3 p + 2i\tau^2 p + \tau^3 p - i\tau^2 p)|p) = i\tau^2 p|p),$$
(III.3.a.iii-9)

$$[K,P]|p) = 2iD|p) = \tau^2 p^2|p) - \tau^2 p^2|p) + 2i\tau p|p) = 2i\tau p|p).$$
(III.3.a.iii-10)

III.3.a.iv *Representations with K diagonal*

Since K is nonlinear it may be more interesting. For K diagonal,

$$K|k) = i\tau^2\frac{d|k)}{d\tau} = k|k).$$ (III.3.a.iv-1)

So

$$i\frac{d|k)}{|k)} = k\frac{d\tau}{\tau^2},$$ (III.3.a.iv-2)

$$ln|k) + C = i\frac{k}{\tau},$$ (III.3.a.iv-3)

$$|k) = B \, exp(i\frac{k}{\tau}),$$ (III.3.a.iv-4)

where B is an arbitrary normalization factor (which may have interesting conditions on it). The effect of K is thus to produce an inversion of τ in its eigenfunctions (perhaps not surprisingly since transversions can be written using inversions, although this is not fully relevant here since there is only one dimension). The other operators give dilatations and translations, as expected, but a transversion involves this inversion (sec. III.1.d, p. 108).

For the other operators

$$P|k) = i\frac{d|k)}{d\tau} = B\frac{k}{\tau^2}exp(i\frac{k}{\tau}),$$ (III.3.a.iv-5)

$$D|k) = -i\tau\frac{d|k)}{d\tau} = -B\frac{k}{\tau}exp(i\frac{k}{\tau}).$$ (III.3.a.iv-6)

Then the commutation relations are

$$[D,P]|k) = iP|k) = (-\frac{k^2}{\tau^3} + 2i\frac{k}{\tau^2} + \frac{k^2}{\tau^3} - i\frac{k}{\tau^2})Bexp(i\frac{k}{\tau})$$

$$= i\frac{k}{\tau^2}Bexp(i\frac{k}{\tau}),$$ (III.3.a.iv-7)

$$[D,K]|k) = -iK|k) = (\frac{-k^2}{\tau} - \frac{-k^2}{\tau} - ik)Bexp(i\frac{k}{\tau}) = -ikBexp(i\frac{k}{\tau}),$$
(III.3.a.iv-8)

$$[K,P]|k) = 2iD|k) = (\frac{k^2}{\tau^2} - 2i\frac{k}{\tau} - \frac{k^2}{\tau^2})Bexp(i\frac{k}{\tau}) = -2i\frac{k}{\tau}Bexp(i\frac{k}{\tau}).$$
(III.3.a.iv-9)

III.3.a.v *Finite transformations of this su(1,1) algebra*

What are the finite transformations for representations with P diagonal? Now

$$P|p) = i\frac{d|p)}{d\tau} = -pexp(ip\tau),$$ (III.3.a.v-1)

gives

$$exp(iaP)|p) = exp\{ip(\tau + a)\},$$ (III.3.a.v-2)

as can be seen from a Taylor expansion around $a = 0$, for which

$$exp(iaP)|p) \Rightarrow exp(ip\tau)\{1 + iap\} = (1 + a\frac{d}{d\tau})exp(ip\tau).$$
(III.3.a.v-3)

Then

$$exp(i\lambda D)exp(ip\tau) \Rightarrow exp(i(1 + \lambda)p\tau),$$ (III.3.a.v-4)

for, expanding around $\lambda = 0$,

$$exp(i(1 + \lambda)p\tau) \Rightarrow exp(ip\tau)\{1 + i\lambda p\tau\} = \{1 + \lambda\tau\frac{d}{d\tau}\}exp(ip\tau),$$
(III.3.a.v-5)

as expected for a dilatation.

For transversions we have finite transformation

$$\tau' = \frac{\tau}{c\tau + 1}, \tag{III.3.a.v-6}$$

$$|p) = exp(ip\tau) \Rightarrow exp(ip\tau') = exp(ip\frac{\tau}{c\tau + 1}) = |p)'. \tag{III.3.a.v-7}$$

Thus on these states

$$K\frac{\tau}{c\tau + 1} = i\tau^2\frac{c\tau + 1 - c\tau}{(c\tau + 1)^2} = i\frac{\tau^2}{(c\tau + 1)^2} = i\tau'^2. \tag{III.3.a.v-8}$$

Expanding around $c = 0$ we get

$$exp(ip\frac{\tau}{c\tau + 1}) \Rightarrow exp(ip\tau)\{1 - icp\tau^2\} = \{1 - c\tau^2\frac{d}{d\tau}\}exp(ip\tau).$$
$$\tag{III.3.a.v-9}$$

What is the effect of the transversion? It maps

$$\tau = \pm\infty \quad to \quad \tau' = \frac{1}{c}, \quad and \quad \tau' = \pm\infty \quad to \quad \tau = -\frac{1}{c}. \tag{III.3.a.v-10}$$

But

$$\tau = 0 \quad is \; mapped \; to \quad \tau' = 0. \tag{III.3.a.v-11}$$

The expectation value (so $p = q$) of P for the transformed state is

$$(|P|) = \int_{-\infty}^{\infty} dq(q|P|p)$$

$$= -p \int d\tau [exp(-iq\frac{\tau}{c\tau + 1})exp(ip\frac{\tau}{c\tau + 1})\{\frac{1}{c\tau + 1} - \frac{c\tau}{(c\tau + 1)^2}\}]$$

$$= -p \int d\tau \frac{1}{(c\tau + 1)^2} = \frac{p}{c}\frac{1}{c\tau + 1}\Big|_{-\infty}^{\infty} = \frac{-2pl}{(1 - c^2l^2)}, \tag{III.3.a.v-12}$$

where the limits $-\infty$ to ∞ have been taken as $-l$ and l. Then in the limit as $l \Rightarrow \infty$,

$$(|P|) = \frac{-2pl}{(1 - c^2l^2)} \Rightarrow 0. \tag{III.3.a.v-13}$$

The expectation value of the momentum between these states, for this realization of this very simple algebra, is 0.

III.3.a.vi *Invariants and their eigenstates*

To give representations we must find labeling invariants, their eigenstates and eigenvalues. What are they for this su(1,1) algebra? Invariants depend only on commutation relations so their expressions in terms of operators are realization-independent. However their realizations come from those of the operators, and this determines their eigenstates and eigenvalues, the basis states and labels of the representations.

For this algebra there is only one invariant,

$$C = D^2 - \frac{1}{2}(PK + KP). \tag{III.3.a.vi-1}$$

From the commutation relations and the realization, the Casimir operator, acting on functions of τ, is realized as

$$C = -(\tau^2 \frac{d^2}{d\tau^2} + \tau \frac{d}{d\tau}) + \frac{1}{2}(\tau^2 \frac{d^2}{d\tau^2} + \tau^2 \frac{d^2}{d\tau^2} + 2\tau \frac{d}{d\tau}) = 0. \tag{III.3.a.vi-2}$$

Thus while for the usual realizations of this algebra there are different values of C giving different representations, here the form of the realization is so different that the Casimir operator does not determine representations, having only a single value, 0. This occurs also for the Euclidean algebra [Mirman (1995c), sec. 2.4, p. 20].

III.3.a.vii *The algebra with internal operators*

Conformal transformations are those of space, and we are quite familiar with operators of its subgroup, the rotation group, including terms that act on internal variables, for it spin. What is the form of the full set of conformal generators when they include internal ones? We start with this realization of su(1,1) analogous to the conformal group. What are the conditions? Of course operators must satisfy the commutation relations. We also assume that space and purely internal parts of operators commute. Operators that satisfy the full conformal algebra commutation relations are known (sec. III.1.e, p. 115), and here we take the simplified version of them to go with this simplified algebra (which does not include the Lorentz operators, since operators we are considering act in only one dimension).
    Then

$$P = i\frac{d}{d\tau}, \tag{III.3.a.vii-1}$$

$$K = i\tau^2 \frac{d}{d\tau} - 2\tau\delta + k, \tag{III.3.a.vii-2}$$

$$D = -i\tau \frac{d}{d\tau} + \delta. \tag{III.3.a.vii-3}$$

These give

$$[D,K] = \tau^2 \frac{d}{d\tau} + 2i\tau\delta - ik = \tau^2 \frac{d}{d\tau} + 2i\tau\delta + [\delta,k], \tag{III.3.a.vii-4}$$

so

$$[\delta,k] = -ik. \tag{III.3.a.vii-5}$$

The same commutation relations hold for the internal part as for the complete operator.

Next

$$[D, P] = iP = -\frac{d}{d\tau} = [-i\tau\frac{d}{d\tau} + \delta, i\frac{d}{d\tau}], \qquad \text{(III.3.a.vii–6)}$$

and, as required,

$$[P, \delta] = 0. \qquad \text{(III.3.a.vii–7)}$$

Finally

$$[K, P] = 2iD = 2i(-i\tau\frac{d}{d\tau} + \delta) = 2\tau\frac{d}{d\tau} + 2i\delta, \qquad \text{(III.3.a.vii–8)}$$

this showing that the form of the $2\tau\delta$ term in K is necessary, and that the term is required. This illustrates an interesting, and unusual, aspect of the nonlinear operators: they contain terms that depend (here as a product) on both coordinates and internal terms. The effect of the internal operator depends on position. Also one internal operator, δ, appears in two generators, K and D.

These assume

$$[P, k] = [P, \delta] = [\tau, k] = [\tau, \delta] = 0. \qquad \text{(III.3.a.vii–9)}$$

The commutation relations give no further conditions on the internal parts of the operators.

For the internal part of the Casimir invariant we get

$$C = D^2 - \frac{1}{2}(PK + KP) = (-i\tau\frac{d}{d\tau} + \delta)^2$$

$$-\frac{1}{2}i\{\frac{d}{d\tau}(i\tau^2\frac{d}{d\tau} - 2\tau\delta + k) + (i\tau^2\frac{d}{d\tau} - 2\tau\delta + k)\frac{d}{d\tau}\}$$

$$= \delta^2 - 2i\delta\tau\frac{d}{d\tau} - \frac{i}{2}(-2\delta + 2k\frac{d}{d\tau} - 4\delta\tau\frac{d}{d\tau}) = \delta^2 + i(\delta + k\frac{d}{d\tau}).$$

$$\text{(III.3.a.vii–10)}$$

While the space part is identically zero, including internal operators gives a Casimir invariant that is not zero, and has a term that acts on space. Thus internal variables allow the possibility of different representations (labeled by the eigenvalues of the invariant).

III.3.a.viii *Representations with internal variables; D diagonal*

How does inclusion of internal variables affect representations? We start with those for which D is diagonal (sec. III.3.a.ii, p. 121). The states are then

$$|n, i) = N_{ni}\tau^n|n, i>, \qquad \text{(III.3.a.viii–1)}$$

given by integers n; $|n, i>$ is the internal part of the state, for which we do not rule out an n dependence and N_{ni} is a normalization factor.

The operators act as

$$D|n, i) = -i\tau \frac{d|n, i)}{d\tau} + \delta|n, i) = -in|n, i) + \sum \delta_{n,ij}|n, j) = d_{n,i}|n, i),$$
(III.3.a.viii–2)

where d is now the dilatation eigenvalue. Since the state is an eigenstate

$$-in|n, i) + \sum \delta_{n,ij}|n, j) = d_{n,i}N_{ni}\tau^n|n, i >.$$ (III.3.a.viii–3)

As states are orthogonal $j = i$, so

$$\delta_{n,ij} = 0, \quad i \neq j,$$
(III.3.a.viii–4)

and the dilatation does not change the internal state. Thus, taking

$$\delta_{n,ij} = \delta_{ni},$$
(III.3.a.viii–5)

we have

$$-in|n, i) + \delta_{ni}|n, i) = -inN_{ni}\tau^n|n, i > +\delta_{ni}N_{ni}\tau^n|n, i >= d_{n,i}|n, i),$$
(III.3.a.viii–6)

$$d_{n,i} = -in + \delta_{ni},$$
(III.3.a.viii–7)

the matrix element of the dilatation, so

$$D|n, i) = (-in + \delta_{ni})|n, i).$$
(III.3.a.viii–8)

For the other operators

$$P|n, i) = i \frac{d|n, i)}{d\tau} = in \frac{N_{ni}}{N_{n-1,i}}|n - 1, i),$$
(III.3.a.viii–9)

leaving the internal part of the state unchanged, and

$$K|n, i) = i\tau^2 \frac{d|n, i)}{d\tau} - 2\tau\delta|n, i) + k|n, i)$$

$$= inN_{ni}\tau^{n+1}|n, i > -2\tau\delta_{ni}|n, i) + k|n, i)$$

$$= (in - 2\delta_{ni}) \frac{N_{ni}}{N_{n+1,i}}|n + 1, i) + \sum k_{n,ij}|n, j).$$ (III.3.a.viii–10)

The action of the momentum is unaffected (except that it is no longer proportional to the eigenvalue of D), while K acquires a matrix element diagonal in the space part, and its off-diagonal matrix elements are changed.

Do commutation relations provide conditions on these states? With

$$[\delta, k] = -ik,$$
(III.3.a.viii–11)

these give

$$[D,P]|n,i) = iP|n,i)$$

$$= ind_{n-1,i}\frac{N_{ni}}{N_{n-1,i}}|n-1,i) - ind_{n,i}\frac{N_{ni}}{N_{n-1,i}}|n-1,i)$$

$$= -in(d_{n,i} - d_{n-1,i})\frac{N_{ni}}{N_{n-1,i}}|n-1,i) = -n\frac{N_{ni}}{N_{n-1,i}}|n-1,i),$$

$$(\text{III.3.a.viii-12})$$

so

$$(d_{n,i} - d_{n-1,i}) = -i, \qquad\qquad (\text{III.3.a.viii-13})$$

which holds for

$$d_{ni} = (-in + \delta_{ni}), \qquad\qquad (\text{III.3.a.viii-14})$$

provided δ_{ni} is independent of n.

Next

$$[D,K]|n,i) = -iK|n,i) = d_{n+1,i}(in - 2\delta_i)\frac{N_{ni}}{N_{n+1,i}}|n+1,i)$$

$$+ \sum d_{n,j}k_{n,ij}|n,j) - d_{ni}(in-2\delta_i)\frac{N_{ni}}{N_{n+1,i}}|n+1,i) - \sum d_{n,i}k_{n,ij}|n,j)$$

$$= (d_{n+1,i} - d_{ni})(in-2\delta_i)\frac{N_{ni}}{N_{n+1,i}}|n+1,i) + \sum(d_{n,j}k_{n,ij} - k_{n,ij}d_{n,i})|n,j)$$

$$= -i(in - 2\delta_i)\frac{N_{ni}}{N_{n+1,i}}|n+1,i) - i\sum k_{n,ij}|n,j), \quad (\text{III.3.a.viii-15})$$

which is satisfied.

Finally

$$[K,P]|n,i) = 2iD|n,i) = 2id_{ni}|n,i) = 2i(-in + \delta_i)|n,i)$$

$$= in\frac{N_{ni}}{N_{n-1,i}}\{(i(n-1) - 2\delta_i)\frac{N_{n-1,i}}{N_{n,i}}|n,i) + \sum k_{n-1,ij}|n-1,j)\}$$

$$- (i(n+1)\frac{N_{n+1,i}}{N_{n,i}})\{(in-2\delta_i)\frac{N_{ni}}{N_{n+1,i}}|n,i) + \sum k_{n,ij}|n,j)\},$$

$$(\text{III.3.a.viii-16})$$

which requires

$$(n+1)\frac{N_{n+1,i}}{N_{n,i}}k_{n,ij} = n\frac{N_{ni}}{N_{n-1,i}}k_{n-1,ij}, \qquad (\text{III.3.a.viii-17})$$

giving the effects of the internal operator and the coordinate part of the basis vector on each other.

III.3.a.ix *Representations with internal operators for P diagonal*

For P diagonal (sec. III.3.a.iii, p. 123),

$$P|p,i) = i\frac{d|p,i)}{d\tau} = -p|p,i) = -p|i > exp(ip\tau), \qquad \text{(III.3.a.ix-1)}$$

$$D|p,i) = -i\tau\frac{d|p,i)}{d\tau} + \delta|p,i) = \tau p|p,i) + \sum \delta_{p,ij}|p,j), \quad \text{(III.3.a.ix-2)}$$

$$K|p,i) = i\tau^2\frac{d|p,i)}{d\tau} - 2\tau\delta|p,i) + k|p,i)$$
$$= -\tau^2 p|p,i) - 2\tau \sum \delta_{p,ij}|p,j) + \sum k_{p,ij}|p,j). \quad \text{(III.3.a.ix-3)}$$

The commutation relations are

$$[D,P]|p,i) = iP|p,i) = -p\{\tau p|p,i) + \sum \delta_{p,ij}|p,i)\}$$
$$+p\{\tau p|p,i) + \sum \delta_{p,ij}|p,i)\} - ip|p,i) = -ip|p,i), \quad \text{(III.3.a.ix-4)}$$

which is satisfied.

Then

$$[D,K]|p,i) = -iK|p,i) = i\tau^2 p|p,i) + 2i\tau \sum \delta_{p,ij}|p,j)$$
$$+ \sum (k_{p,ij}\delta_{p,jk} - \delta_{p,ij}k_{p,jk})|p,k), \quad \text{(III.3.a.ix-5)}$$

and for this realization also

$$[\delta,k] = -ik. \qquad \text{(III.3.a.ix-6)}$$

The last one is

$$[K,P]|p,i) = 2iD|p,i) = 2i\{\tau p|p,i) + \sum \delta_{p,ij}|p,j)\}, \quad \text{(III.3.a.ix-7)}$$

imposing no conditions.

III.3.a.x *K diagonal representations with internal operators*

And for K diagonal (sec. III.3.a.iv, p. 124),

$$K|\kappa,i) = i\tau^2\frac{d|\kappa,i)}{d\tau} - 2\tau\delta|\kappa,i) + k|\kappa,i) = \kappa_i|\kappa,i). \qquad \text{(III.3.a.x-1)}$$

So, assuming that the derivative has no effect on the internal label,

$$i\tau^2\frac{d|\kappa,i)}{d\tau} = \kappa_i|\kappa,i) + 2\tau \sum \Delta(\kappa,ij)|\kappa,j) - \sum k_{\kappa,ij}|\kappa,j), \quad \text{(III.3.a.x-2)}$$

where Δ is the matrix element of δ. Since by orthogonality the derivative and τ do not change the internal state label and the last two terms cannot cancel because their τ dependence differs, we get

$$i\frac{d|\kappa,i)}{|\kappa,i)} = (\kappa_i + 2\tau\Delta(\kappa,i) - k_i)\frac{d\tau}{\tau^2}, \qquad \text{(III.3.a.x-3)}$$

$$ln|\kappa,i) + C = i\frac{\kappa_i - k_i}{\tau} - 2i\Delta(\kappa,i)ln(\tau), \qquad \text{(III.3.a.x-4)}$$

$$\frac{|\kappa,i)}{\tau^{-2i\Delta(\kappa,i)}} = B_{\kappa i}exp(i\frac{\kappa_i - k_i}{\tau}), \qquad \text{(III.3.a.x-5)}$$

$$|\kappa,i) = B_{\kappa i}\tau^{-2i\Delta(\kappa,i)}exp(i\frac{\kappa_i - k_i}{\tau}). \qquad \text{(III.3.a.x-6)}$$

If there is an internal dependence it is only through k_i, the matrix element of k, $\Delta(\kappa,i)$, and the normalization. The internal operators produce additional space dependence.

Then

$$P|\kappa,i) = i\frac{d|\kappa,i)}{d\tau} = B_{\kappa i}\tau^{-2i\Delta(\kappa,i)}(\frac{\kappa_i - k_i}{\tau^2})exp(i\frac{\kappa_i - k_i}{\tau})$$

$$+2\Delta(\kappa,i)B_{\kappa i}\tau^{-2i\Delta(\kappa,i)-1}exp(i\frac{\kappa_i-k_i}{\tau}) = (\frac{\kappa_i - k_i}{\tau^2} + \frac{2\Delta(\kappa,i)}{\tau})|\kappa,i),$$

$$\text{(III.3.a.x-7)}$$

$$D|\kappa,i) = -i\tau\frac{d|\kappa,i)}{d\tau} + \Delta(\kappa,i)|\kappa,i) = -B_{\kappa i}\tau^{-2i\Delta(\kappa,i)}\frac{\kappa_i-k_i}{\tau}exp(i\frac{\kappa_i-k_i}{\tau})$$

$$+\Delta(\kappa,i)|\kappa,i) - 2\Delta(\kappa,i)B_{\kappa i}\tau^{-2i\Delta(\kappa,i)}exp(i\frac{\kappa_i - k_i}{\tau})$$

$$= \{-\frac{\kappa_i - k_i}{\tau} - \Delta(\kappa,i)\}|\kappa,i). \qquad \text{(III.3.a.x-8)}$$

The eigenvector of K is changed, so is no longer an eigenvector; there is additional space dependence. Something like this might be expected since a transversion can be interpreted as a space-dependent dilatation (sec. I.5, p. 43).

From the commutation relations we have

$$[D,K]|\kappa,i) = -iK|\kappa,i) = -i\kappa_i|\kappa,i) = \kappa_i\{\frac{-(\kappa_i - k_i)}{\tau} - \Delta(\kappa,i)\}|\kappa,i)$$

$$-\kappa_i\{\frac{-(\kappa_i - k_i)}{\tau} - \Delta(\kappa,i)\}|\kappa,i) - i(\kappa_i - k_i)|\kappa,i), \qquad \text{(III.3.a.x-9)}$$

requiring $k_i = 0$, so that this internal operator does not appear in this type of representation, of this very simple algebra. And

$$[K,P]|\kappa,i) = 2iD|\kappa,i) = -2i\frac{\kappa_i}{\tau}|\kappa,i)$$

$$= \{\frac{-2i\kappa_i}{\tau} - 2i\Delta(\kappa,i) + \kappa_i(\frac{\kappa_i}{\tau^2} + \frac{2\Delta(\kappa,i)}{\tau})\}|\kappa,i)$$

$$-\kappa_i\{\frac{\kappa_i}{\tau^2} + \frac{2\Delta(\kappa,i)}{\tau}\}|\kappa,i), \qquad \text{(III.3.a.x-10)}$$

which gives $\delta = 0$, so for this type of representation there are no internal operators. Then

$$[D, P]|\kappa, i) = iP|\kappa, i) = i\frac{\kappa_i}{\tau^2}|\kappa, i)$$

$$= \{-\frac{\kappa_i}{\tau}\frac{\kappa_i}{\tau^2} + 2i\frac{\kappa_i}{\tau^2} + \frac{\kappa_i}{\tau}\frac{\kappa_i}{\tau^2} - i\frac{\kappa_i}{\tau^2}\}|\kappa, i), \qquad \text{(III.3.a.x–11)}$$

imposing no conditions.

III.3.b   Relating SU(1,1) generators to those of the simple conformal group

The conformal group is noncompact so we are considering the noncompact extension of the nonlinear, inhomogeneous realizations of the su(2) algebra, su(1,1), which is isomorphic to so(2,1).

Group SO(2,1) is smaller, so there are fewer generators for this analog of the conformal group, three (sec. III.3.a, p. 120). We write

$$[D, Q] = iQ, \qquad \text{(III.3.b–1)}$$

$$[D, K] = -iK, \qquad \text{(III.3.b–2)}$$

$$[K, Q] = 2iD, \qquad \text{(III.3.b–3)}$$

with the label of the momentum generator changed, and we wish to relate these to the su(1,1) operators. They are taken as

$$K = (L_{13} + L_{23}) = L_+, \qquad \text{(III.3.b–4)}$$

$$Q = (L_{13} - L_{23}) = L_-, \qquad \text{(III.3.b–5)}$$

$$D = L_{12}, \qquad \text{(III.3.b–6)}$$

and P is not one of them, but is defined as

$$P = L_{12} + L_{13} = D + \frac{1}{2}(L_+ + L_-) = D + \frac{1}{2}(K + Q), \qquad \text{(III.3.b–7)}$$

$$Q = 2(P - D) - K. \qquad \text{(III.3.b–8)}$$

From the commutation relations (sec. III.1.d.v, p. 112), for one dimension, to relate these to the generators of su(1,1), forming the same algebra as that of so(2,1), with

$$g_{11} = -1, g_{22} = -1, g_{33} = 1, \qquad \text{(III.3.b–9)}$$

we have

$$[D, Q] = iQ = [L_{12}, L_-] = iL_-, \qquad \text{(III.3.b–10)}$$

$$[D, K] = -iK = [L_{12}, L_+] = -iL_+, \qquad \text{(III.3.b–11)}$$

$$[K, Q] = 2iD = [L_+, L_-] = 2iL_{12}, \qquad \text{(III.3.b-12)}$$

which verifies that these three operators form an su(1,1) algebra.

Hence

$$[D, P] = [L_{12}, L_{12} + L_{13}] = [L_{12}, L_{13}] = \frac{1}{2}[L_{12}, L_+ + L_-]$$

$$= -\frac{1}{2}i(L_+ - L_-) = -\frac{1}{2}iL_{23} = -i(K - Q). \quad \text{(III.3.b-13)}$$

$$[K, P] = [L_+, L_{12} + L_{13}] = iL_+ + \frac{1}{2}[L_+, L_+ + L_-]$$

$$= iL_+ + iL_{12} = i(D + K). \qquad \text{(III.3.b-14)}$$

Below (sec. III.4.e, p. 154) we consider representations in which P — a sum of su(1,1) generators — is diagonal, an unusual aspect. This leads to K being nonlinear.

III.3.b.i   *The Lorentz subgroup* SE(2)

The conformal group is obtained from ones defined over a space of six real dimensions, but is realized over a space of four real dimensions. There is a smaller, and more familiar, group that that models this, the Lorentz group. For massless objects the little group is SE(2), the Euclidean group in two real dimensions, a subgroup of the Lorentz group [Mirman (1995c), sec. 2.4, p. 20].

For this there is rotation operator M, for rotations around the axis of propagation, which corresponds to the set of algebra generators, $M_{\mu\nu}$; also there are two momentum operators P_1 and P_2, mixed by M. Adjoining the two operators R_1 and R_2, which correspond to the transversion generators, the K's, and the boost B, which corresponds to dilatation D, we get an algebra that serves as a model of that of the conformal group, as can be seen from the commutation relations [Mirman (1995c), sec. 2.3, p. 15].

However the realization of this algebra, when used as a little group for massless objects for the Poincaré group [Mirman (1995c), sec. 2.2.3, p. 14] is different than the realization for the algebra of the conformal group. The N's are realized homogeneously, rather than inhomogeneously as for the P's of the conformal group, and the R's are likewise realized homogeneously, rather than nonlinearly, as for the K's.

They are so realized because the operators are written in terms of the 3+1 coordinates on which the Lorentz group acts, rather than as operators over a two-dimensional space, which is that of SE(2). But of course, they can be written in terms of just two coordinates, and then would be realized in the same way as the generators of the conformal group.

III.3.b.ii *The su(2) cover of so(3)*

It is useful to also review the SU(2) cover of SO(3) [Hamermesh (1962), p. 348; Mirman (1995a), sec. X.6, p. 287]. The SU(2) algebra acts on complex variables ζ, η, and that of SO(3) on real ones x, y, z. These are related by

$$x = \frac{1}{2}(\eta^2 - \zeta^2), \quad y = \frac{1}{2i}(\eta^2 + \zeta^2), \quad z = \eta\zeta. \qquad \text{(III.3.b.ii–1)}$$

The algebra operators for SO(3) are

$$L_z = i(x\frac{d}{dy} - y\frac{d}{dx}), \qquad \text{(III.3.b.ii–2)}$$

$$L_y = i(z\frac{d}{dx} - x\frac{d}{dz}), \qquad \text{(III.3.b.ii–3)}$$

$$L_x = i(y\frac{d}{dz} - z\frac{d}{dy}), \qquad \text{(III.3.b.ii–4)}$$

and, for SU(2),

$$J_z = i(\eta\frac{d}{d\eta} - \zeta\frac{d}{d\zeta}), \qquad \text{(III.3.b.ii–5)}$$

$$J_+ = i\eta\frac{d}{d\zeta}, \quad J_- = i\zeta\frac{d}{d\eta}. \qquad \text{(III.3.b.ii–6)}$$

$$J_+ = \frac{1}{2}(J_x + iJ_y), \quad J_- = \frac{1}{2}(J_x - iJ_y). \qquad \text{(III.3.b.ii–7)}$$

So

$$J_x = J_+ + J_- = i(\eta\frac{d}{d\zeta} + \zeta\frac{d}{d\eta}), \qquad \text{(III.3.b.ii–8)}$$

$$J_y = -i(J_+ - J_-) = (\eta\frac{d}{d\zeta} - \zeta\frac{d}{d\eta}). \qquad \text{(III.3.b.ii–9)}$$

Now

$$\frac{d}{dx} = \frac{1}{\eta}\frac{d}{d\eta} - \frac{1}{\zeta}\frac{d}{d\zeta}, \quad \frac{d}{dy} = \frac{i}{\eta}\frac{d}{d\eta} + \frac{i}{\zeta}\frac{d}{d\zeta}, \quad \frac{d}{dz} = \frac{1}{\zeta}\frac{d}{d\eta} + \frac{1}{\eta}\frac{d}{d\zeta}.$$
$$\text{(III.3.b.ii–10)}$$

Therefore

$$L_z = i(x\frac{d}{dy} - y\frac{d}{dx}) = -(\eta\frac{d}{d\eta} - \zeta\frac{d}{d\zeta}), \qquad \text{(III.3.b.ii–11)}$$

$$L_y = i(z\frac{d}{dx} - x\frac{d}{dz}) = \frac{3i}{2}(\zeta\frac{d}{d\eta} - \eta\frac{d}{d\zeta}) - \frac{i}{2}(\frac{\eta^2}{\zeta}\frac{d}{d\eta} - \frac{\zeta^2}{\eta}\frac{d}{d\zeta}), \quad \text{(III.3.b.ii–12)}$$

$$L_x = i(y\frac{d}{dz} - z\frac{d}{dy}) = \frac{3}{2}(\zeta\frac{d}{d\eta} + \eta\frac{d}{d\zeta}) + \frac{1}{2}(\frac{\eta^2}{\zeta}\frac{d}{d\eta} + \frac{\zeta^2}{\eta}\frac{d}{d\zeta}). \quad \text{(III.3.b.ii–13)}$$

While these operators are of course linear over their space of definition, they are nonlinear over a smaller space (sec. III.5.b, p. 162).

III.3.b.iii   SO(2,1) *representations*

The SO(2,1) algebra is found from that of SO(3) by setting $x_3 \Rightarrow ix_3$, giving, instead of commutation relations,

$$[L_{12},L_+] = iL_+, \quad [L_{12},L_-] = -iL_-, \quad [L_+,L_-] = 2iL_{12}, \qquad \text{(III.3.b.iii-1)}$$

the set

$$[L_{12},L_+] = -iL_+, \quad [L_{12},L_-] = iL_-, \quad [L_+,L_-] = 2iL_{12}, \qquad \text{(III.3.b.iii-2)}$$

so that $L_\pm$ are antihermitian instead of hermitian. The SO(3) invariant

$$L^2 = L_{12}^2 + \frac{1}{2}(L_+L_- + L_-L_+), \qquad \text{(III.3.b.iii-3)}$$

becomes

$$L^2 = L_{12}^2 - \frac{1}{2}(L_+L_- + L_-L_+). \qquad \text{(III.3.b.iii-4)}$$

This being an absolute square is always positive, as is L_{12}^2, so that here, instead of a maximum value of m, there is a minimum. Thus the states of the representation go from l (the representation label) to ∞, and $-l$ to $-\infty$. In the matrix elements appears label m, whose magnitude instead of always being smaller than or equal to l, is equal to it, or larger. This results in a minus sign, and so an i in the square root, making the operators antihermitian.

Matrix elements of SO(2,1) generators, for representations in which the usual generators are diagonal, are known [Maekawa (1979a,b,1980)], say by continuation from those of SO(3).

III.3.b.iv   *The representations of the SO(2,1) group*

To find explicit group representations we start with the spherical harmonics and substitute

$$exp(i\phi) \Rightarrow exp(i\phi), \quad cos\theta \Rightarrow cosh\theta, \quad sin\theta \Rightarrow sinh\theta. \qquad \text{(III.3.b.iv-1)}$$

This also gives the explicit form of the generators of this algebra.

For SO(3), that L^2 cannot be negative requires that there be a largest and smallest value of m, and since the generators are hermitian, these must be negatives of each other, and also differ by an integer. Thus l has to be an integer (for SU(2), a half-integer). However for SO(2,1), the bounds are the smallest positive and negative integer state labels (rather than the largest). These then are not connected by step operators, so are not required to differ by an integer. Representation label l therefore need not be an integer, but can have any real value. Also $(sin\theta)^l$ must be real for all values of θ, which limits l to integers, but $(sinh\theta)^l$ is real for all positive values of θ, so there is no limit on l — another way of seeing that it can have any real value. But the labels of the states, for each representation, differ by integers (except for the extreme, boundary, values).

III.3.b.v *The labels of the* SO(2,1), SU(1,1) *groups*

The representations of these groups are well-known [Bargmann (1947), p. 569, 609]. There are two labels, q, the eigenvalues of the Casimir operator C, which labels the representations, and the eigenvalues, m, of L_{12}, which labels the states.

Then

C_q^0: where q is any positive number, and m has all integral values $0, \pm 1, \pm 2, \ldots$;

$C_q^{\frac{1}{2}}$: q is any number such that $\frac{1}{4} < q < \infty$, and m runs over all half-integral values, $\pm \frac{1}{2}, \pm \frac{3}{2}, \ldots$;

D_k^+: where k is one of the numbers $\frac{1}{2}, 1, \frac{3}{2}, \ldots$, $q = k(1 - k)$, so has the values $\frac{1}{4}, 0, -\frac{3}{4}, \ldots$. And m runs over all $k, k + 1, k + 2, \ldots$;

D_k^-: again k is one of $\frac{1}{2}, 1, \frac{3}{2}, \ldots$, $q = k(1 - k)$, so has the values $\frac{3}{4}, 0, -\frac{1}{4}, \ldots$. Now m runs over all values $-k, -(k + 1), -(k + 2), \ldots$.

The two C classes are continuous as q runs over a continuous interval, going to infinity, thus the set of representations in the classes is continuous, while the D classes are discrete, as q can take only discrete values, so the set of these representations is discrete. However in each representation, for all classes, the set of states is discrete. These representations are all infinite-dimensional.

III.4   The other groups with fifteen parameters

There are, besides the conformal group, two others with fifteen parameters, SO(4,2) and SU(2,2), (plus complex extensions). Studying relationships among these helps in finding, and understanding, representations of the conformal group. We consider this next.

III.4.a   The SO(4,2) realization

The conformal group algebra is isomorphic to that of group SO(4,2), and to that of SU(2,2) [Binegar, Fronsdal and Heidenreich (1983); Carruthers (1971); Mack (1977); Mack and Salam (1969)]. This we start to explore here.

III.4.a.i   *The SU(2,2) cover of SO(4,2)*

Group SO(4,2) is covered by SU(2,2) since their algebras are isomorphic; the groups then are homomorphic. Since we are not investigating the global properties of these groups, only their algebras, we do not consider this explicitly but just note it.

III.4.a.ii *How the algebras are related*

The conformal group acts on a space of four real variables, with metric (+,+,+,-) or (-,-,-,+), while SO(4,2) acts on one of six real ones, with metric (+,+,+,-,+,-) or (-,-,-,+,-,+). How are coordinates of these spaces related? What do we learn from this relationship, what are its implications?

We start by finding the relationship between the algebras so(4,2) and that of the conformal group, giving two different types of expressions, showing that the generators of the two algebras can be expressed as linear sums of each other and that the commutation relations are satisfied using either the conformal generators or those of so(4,2), so that they are isomorphic.

III.4.a.iii *Expressing generators of one group in terms of the other*

Representations of the Poincaré group are (usually) ones with the momentum operators diagonal. Since the relevant representations of the conformal group are those that are also representations of the Poincaré subgroup we have to find conformal representations having diagonal momentum generators, an aspect considered below (sec. III.4.e, p. 154). Here conformal generators are written in terms of those of algebra so(4,2), so it is necessary to diagonalize (generators of) transformations that are sums of a rotation and of a boost, of the form of operators N_i [Mirman (1995c), sec. 2.3.1, p. 15]. The momentum operators, with metric (+,+,+,-,+,-) or equivalently (-,-,-,+,-,+), are P_μ. We have to relate these generators to the so(4,2) operators, L. The commutation relations for all orthogonal algebras are the same except for the ranges of indices, and (up to relabeling), as can be seen from the commutation relations (sec. III.1.d.v, p. 112), the three generators that commute with each other are L_{12}, L_{34}, L_{56}. None of the others commute with all these three commuting labeling operators.

Generators of so(4,2) obey commutation relations

$$[L_{\kappa\lambda}, L_{\mu\nu}] = i(g_{\lambda\mu}L_{\kappa\nu} - g_{\kappa\mu}L_{\lambda\nu} - g_{\lambda\nu}L_{\kappa\mu} + g_{\kappa\nu}L_{\lambda\mu}); \qquad \text{(III.4.a.iii–1)}$$

indices run from 1 to 6, and the metric is (+,+,+,-,+,-). These are related to the generators of the conformal group, with the Lorentz generators denoted by $M_{\mu\nu}$, $\mu = 1, \ldots, 4$ [Carruthers (1971)],

$$M_{\beta\gamma} = L_{\beta\gamma}, \qquad \text{(III.4.a.iii–2)}$$

$$P_\mu = (L_{5\mu} + L_{6\mu}), \qquad \text{(III.4.a.iii–3)}$$

$$K_\mu = (L_{5\mu} - L_{6\mu}), \qquad \text{(III.4.a.iii–4)}$$

$$D = L_{65}, \qquad \text{(III.4.a.iii–5)}$$

so

$$L_{5\mu} = \frac{1}{2}(P_\mu + K_\mu), \qquad \text{(III.4.a.iii–6)}$$

$$L_{6\mu} = \frac{1}{2}(P_\mu - K_\mu). \qquad \text{(III.4.a.iii-7)}$$

And with metric $(+,-,-,-)$ Poincaré generators have commutation relations,

$$[M_{\kappa\lambda}, M_{\mu\nu}] = i(g_{\lambda\mu}M_{\kappa\nu} - g_{\kappa\mu}M_{\lambda\nu} - g_{\lambda\nu}M_{\kappa\mu} + g_{\kappa\nu}M_{\lambda\mu}), \qquad \text{(III.4.a.iii-8)}$$

$$[P_\lambda, M_{\mu\nu}] = i(g_{\lambda\mu}P_\nu - g_{\lambda\nu}P_\mu), \qquad \text{(III.4.a.iii-9)}$$

$$[P_\mu, P_\nu] = 0. \qquad \text{(III.4.a.iii-10)}$$

The remaining commutation relations are

$$[D, P_\mu] = iP_\mu, \quad [D, M_{\mu\nu}] = 0, \qquad \text{(III.4.a.iii-11)}$$

$$[D, K_\mu] = -iK_\mu, \quad [K_\mu, K_\nu] = 0, \qquad \text{(III.4.a.iii-12)}$$

$$[K_\mu, P_\nu] = 2i(g_{\mu\nu}D - M_{\mu\nu}), \qquad \text{(III.4.a.iii-13)}$$

where $g_{\mu\nu}$ is the metric,

$$[K_\lambda, M_{\mu\nu}] = i(g_{\lambda\mu}K_\nu - g_{\lambda\nu}K_\mu). \qquad \text{(III.4.a.iii-14)}$$

These relate the generators of the two groups.

III.4.a.iv   *The conformal algebra and that of so(4,2) are isomorphic*

We now have to show that the algebra so(4,2) is isomorphic to that of the conformal group [Carruthers (1971), p. 6; Mack (1977), p. 5], thus have to check the commutation relations. The Lorentz generators clearly have the correct commutation relations since they are the same as those of the L's, except for the range of indices. The other commutation relations have to be verified.

Since the P's are inhomogeneous operators they must commute. Thus, using the signature of the metric, and $\mu \neq \nu$,

$$P_\mu P_\nu - P_\nu P_\mu = (L_{5\mu} - L_{6\mu})(L_{5\nu} - L_{6\nu}) - (L_{5\nu} - L_{6\nu})(L_{5\mu} - L_{6\mu})$$
$$= L_{5\mu}L_{5\nu} - L_{5\mu}L_{6\nu} - L_{6\mu}L_{5\nu} + L_{6\mu}L_{6\nu} - L_{5\nu}L_{5\mu} + L_{5\nu}L_{6\mu} + L_{6\nu}L_{5\mu} - L_{6\nu}L_{6\mu}$$
$$= [L_{5\mu}, L_{5\nu}] + [L_{6\mu}, L_{6\nu}] = 0. \qquad \text{(III.4.a.iv-1)}$$

The argument is the same for the K's, since the sign change does not affect it, and these commute. Note that the P's commute for this algebra but not for its compact extension (sec. III.1.f, p. 115).

Now, using the metric,

$$[K_\mu, P_\nu] = [L_{5\mu}, L_{5\nu}] - [L_{6\mu}, L_{6\nu}] = -2iL_{\mu\nu} = -2iM_{\mu\nu}, \quad \mu \neq \nu,$$
$$\text{(III.4.a.iv-2)}$$
$$[K_\mu, P_\mu] = 2[L_{5\mu}, L_{6\mu}] = 2ig_{\mu\mu}L_{65} = 2ig_{\mu\mu}D, \qquad \text{(III.4.a.iv-3)}$$
$$[D, K_\mu] = [L_{65}, L_{5\mu} - L_{6\mu}] = i(L_{6\mu} - L_{5\mu}) = -iK_\mu, \qquad \text{(III.4.a.iv-4)}$$

$$[D, P_\mu] = [L_{65}, L_{5\mu} + L_{6\mu}] = i(L_{6\mu} + L_{5\mu}) = iP_\mu, \qquad \text{(III.4.a.iv-5)}$$

so the operators of the so(4,2) algebra and that of the conformal group obey the same commutation relations. These two algebras are thus isomorphic. The need for the metric emphasizes that the algebra to which the conformal algebra is isomorphic must be noncompact.

For comparison, the N operators [Mirman (1995c), sec. 2.3.1, p. 15] are

$$N_1 = -i(t\frac{d}{dx_1} + x_1\frac{d}{dt} - z\frac{d}{dx_1} + x_1\frac{d}{dz}), \qquad \text{(III.4.a.iv-6)}$$

$$N_2 = -i(t\frac{d}{dx_2} + x_2\frac{d}{dt} - z\frac{d}{dx_2} + x_2\frac{d}{dz}). \qquad \text{(III.4.a.iv-7)}$$

So

$$N_1 N_2 = -x_1\frac{d}{dx_2} + x_1\frac{d}{dx_2} = 0. \qquad \text{(III.4.a.iv-8)}$$

Thus since their product is zero, they commute.

The part $(-z\frac{d}{dx_1} + x_1\frac{d}{dz})$ is a rotation in the x, z plane (around y), while $(t\frac{d}{dx_1} + x_1\frac{d}{dt})$ is a pseudo-rotation in the x, t plane, a boost. Operators $(L_{5\mu} - L_{6\mu})$ correspond to this. Note that $(L_{5i} - L_{6i})$, $i = 1,2,3$, are sums of a rotation and a boost, while $(L_{54} - L_{64})$ is a sum of a boost and a rotation. Thus all are of the same form as the N's; diagonalization of the P's is somewhat unusual for they are sums of operators of different types, rotations and boosts.

A special conformal transformation on state $|s\rangle$ gives,

$$exp(i\eta K_\mu)|s\rangle = exp[-i\eta(L_{5\mu} + L_{6\mu})]|s\rangle. \qquad \text{(III.4.a.iv-9)}$$

Generator K is nonlinear for the conformal group, but in this space it is a linear transformation, although on more variables.

III.4.a.v   *A realization using Dirac matrices*

There is another realization of the algebra which we note [Carruthers (1971), p. 6]. It is

$$M_{\mu\nu} \Rightarrow \frac{1}{2}\sigma_{\mu\nu} = -\frac{i}{4}[\gamma_\mu, \gamma_\nu], \quad D \Rightarrow -\frac{1}{2}\gamma_5, \qquad \text{(III.4.a.v-1)}$$

$$P_\mu \Rightarrow \frac{1}{2}\gamma_\mu(1 - i\gamma_5), \quad K_\mu \Rightarrow \frac{1}{2}\gamma_\mu(1 + i\gamma_5). \qquad \text{(III.4.a.v-2)}$$

Here the fifteen operators are written in terms of 5 Dirac matrices and the unit matrix. These can be taken as a conformal algebra realization on the spin states (for spin-$\frac{1}{2}$).

To show that these obey the conformal algebra commutation relations we use [Schweber (1962), p. 69]

$$\{\gamma_\mu, \gamma_\nu\} = \gamma_\mu\gamma_\nu + \gamma_\nu\gamma_\mu = 2g_{\mu\nu}I, \qquad \text{(III.4.a.v-3)}$$

where g is the metric and I the unit matrix. Thus

$$y_4^2 = 1, \quad y_i^2 = -1. \tag{III.4.a.v-4}$$

Also

$$y_5 = y_4 y_1 y_2 y_3, \quad y_5^2 = -1. \tag{III.4.a.v-5}$$

Thus

$$y_4 y_5 = -y_5 y_4 = y_1 y_2 y_3, \tag{III.4.a.v-6}$$

$$y_1 y_5 = -y_5 y_1 = y_4 y_2 y_3, \tag{III.4.a.v-7}$$

$$y_2 y_5 = -y_5 y_2 = -y_4 y_1 y_3, \tag{III.4.a.v-8}$$

$$y_3 y_5 = -y_5 y_3 = y_4 y_1 y_2. \tag{III.4.a.v-9}$$

So for the commutators

$$[y_\mu, y_5 y_\nu] = 0, \quad \mu \neq \nu, \tag{III.4.a.v-10}$$

which is true for all four positions of y_5 in the commutator, and

$$[y_\mu y_5, y_\nu y_5] = [y_\mu, y_\nu]. \tag{III.4.a.v-11}$$

Then

$$[P_\mu, P_\nu] = [K_\mu, K_\nu] \sim [y_\mu, y_\nu] - [y_\mu y_5, y_\nu y_5] = 0. \tag{III.4.a.v-12}$$

That $\sigma_{\mu\nu}$ forms the Lorentz algebra is clear. Also

$$[D, P_\mu] = -\frac{1}{4}([y_5, y_\mu] - i[y_5, y_\mu y_5]) = iP_\mu, \tag{III.4.a.v-13}$$

and similarly for K_μ. Finally

$$[K_\mu, P_\nu] = 2i(g_{\mu\nu}D - M_{\mu\nu}) = \frac{1}{4}([y_\mu, y_\nu] + [y_\mu y_5, y_\nu y_5]) = \frac{1}{2}[y_\mu, y_\nu]$$
$$= -2i\sigma_{\mu\nu}, \quad \mu \neq \nu, \tag{III.4.a.v-14}$$

$$[K_\mu, P_\mu] = -\frac{1}{2}i[y_5 y_\mu, y_\mu] = \frac{i}{2}(y_\mu y_5 y_\mu - y_5 y_\mu y_\mu) = -ig_{\mu\mu}y_5 = 2ig_{\mu\mu}D. \tag{III.4.a.v-15}$$

Thus they do obey the conformal algebra commutation relations. These 15 operators close on themselves under commutation so are a realization of the so(4,2) algebra, which is also demonstrated by the expressions for the orthogonal generators in terms of the conformal ones (sec. III.4.a.iii, p. 138).

III.4.b   The su(3,1) algebra and its implications

There is another algebra with 15 generators, su(3,1). Whether there
can be physical connections among these is not clear, but it is worth
mentioning some possibilities.

Our universe has objects with integer spin, bosons, and also ones
with half-(odd)-integer spin, fermions. Can there be groups whose representations contain both? In fact it is quite familiar that there are. For
example the octet representation of su(3) has objects with integer (iso-)
spin and others with half-integer spin. This is also the adjoint (regu-
lar) representation [Mirman (1995a), sec. XIV.1.d, p. 405] to which the
generators belong, and these mix the two types of basis states.

This algebra is not regarded as giving a set of transformations, but
rather is taken as a spectrum-generating algebra: its spectrum (the set
of its basis states) corresponds to physical states. But we might regard
such algebras in different ways.

That subalgebra transforming according to the su(2) vector representation can be taken to act over space (so isomorphic to the so(3)
algebra), thus its basis states would be spin basis vectors. Generators
that mix fermions and bosons would not be hermitian, so the finite
transformations would not be unitary. However the su(3) octet has two
such sets of operators, both transforming as su(2) spinors, and hermitian conjugates of each other.

One interpretation of them would be as creation and annihilation
operators. That they are not hermitian, and the finite transformations
not unitary, is not only acceptable, but necessary. They change the
number of particles, so probability.

Some of these operators create (and annihilate) fermions, others
bosons [Mirman (1995b), sec. 8.1.3, p. 149], so what are their commutation relations? The defining su(2) representation [Mirman (1995a),
sec. XIV.1.c, p. 404] goes with spin-$\frac{1}{2}$, and the creation and annihilation operators for these must anticommute [Mirman (1995b), chap. 8,
p. 146]. This is true for all half-integer spins (which transform as products of odd numbers of defining-representation basis vectors). Operators for integral spin transform as products of (an even number of)
spin-$\frac{1}{2}$ basis vectors so are taken to have the same commutation relations. If a, a^* anticommute, then a, aa^*, as well as aa^*, a^*a, commute.
Thus creation and annihilation operators of bosons commute.

Generators of the su(2) subalgebra transforming according to its adjoint representation of course can also be realized in terms of creation
and annihilation operators, as products. Their (well-known) commutation relations then follow from those of the creation and annihilation
operators.

Why must transformations be unitary — in many cases? A rotation
operator correlates basis vectors defined using one set of axes with
those defined with respect to another, that is rotates them. If such cor-

relation also changed the normalization (probability) the consistency of the formalism, both mathematically and physically, would be questionable.

A rotation operator cannot rotate an object, but merely correlates basis vectors. Such an actual transformation is performed by momentum operators containing nonlinear terms (giving rotated basis vectors from unrotated ones). These are products of operators that need not, and likely could not, be hermitian. An operator that annihilates a particle is multiplied by ones that create others, so the total probability is conserved. Unitarity is only one reason, conservation of particle type, of charge, and so on provide others. The absorption of a photon by an electron causes the electron to rotate. But it cannot turn one electron into two.

Groups then that have nonhermitian generators, and whose representations include both fermions and bosons, are physically not only reasonable, expected and known, but necessary.

What is interesting about su(3,1) is that its su(3) subrepresentations do contain both fermions and bosons, while representations of su(3,1) form infinite towers of su(3) representations, as the representations of the Lorentz group form such towers of su(2) representations (sec. A.7, p. 242). The relevance of this is not clear since the arguments for the relevance of the conformal realization to other algebras do not immediately carry over to su(3,1). However the possibility of such connection is interesting.

III.4.c   Relating coordinates of four and of six dimensions

There are thus relationships between the Lie algebra generators of these 15 parameter groups. They act on different sets of coordinates, of four and of six real dimensions. How are the four real coordinates of our space related to the six real ones of SO(4,2) [Dirac (1936); Mack and Salam (1969), p. 192]? Does it have meaning to so relate them?

With the x's the coordinates (one imaginary) of our four-dimensional space on which the Poincaré group acts, define

$$r^2 = x_1^2 + \ldots + x_4^2, \tag{III.4.c-1}$$

$$x_5 = \frac{(1-r^2)}{2}, \quad x_6 = \frac{(1+r^2)}{2}, \tag{III.4.c-2}$$

so

$$x_5 + x_6 = 1. \tag{III.4.c-3}$$

Also

$$x_1^2 + \ldots + x_4^2 + x_5^2 = r^2 + \frac{1}{4}(1 - 2r^2 + r^4) = \frac{1}{4}(1 + 2r^2 + r^4) = x_6^2, \tag{III.4.c-4}$$

giving coordinates for the space of SO(4,2).

Generators of SO(4,2)

$$L_{ij} = i(x_i \frac{d}{dx_j} - x_j \frac{d}{dx_i}), \quad i, j = 1, \ldots, 6, \qquad \text{(III.4.c-5)}$$

$$L_{i(4,6)} = -L_{(4,6)i} = i(x_i \frac{d}{idx_{4,6}} - ix_{4,6}\frac{d}{dx_i}) = (x_i \frac{d}{dx_{4,6}} + x_{4,6}\frac{d}{dx_i}),$$
$$i \neq 4, 6, \quad \text{(III.4.c-6)}$$

(sec. III.1.f, p. 115) include, $j = 1, 2, 3,$

$$L_{5,6\mu} = i(x_{5,6}\frac{d}{dx_\mu} - x_\mu \frac{d}{dx_{5,6}}) = i(\frac{1}{2}(1 \mp r^2)\frac{d}{dx_\mu} - 2x_\mu \frac{d}{d(1 \mp r^2)})$$
$$= -i(\mp 2x_\mu \frac{d}{d(r^2)} - \frac{1}{2}(1 \mp r^2)\frac{d}{dx_\mu}), \quad \text{(III.4.c-7)}$$

$$L_{5j} = i(x_5 \frac{d}{dx_j} - x_j \frac{d}{dx_5}) = -i(2x_j \frac{d}{d(1 - r^2)} - \frac{1}{2}(1 - r^2)\frac{d}{dx_j})$$
$$= i(2x_j \frac{d}{d(r^2)} + \frac{1}{2}(1 - r^2)\frac{d}{dx_j}), \quad \text{(III.4.c-8)}$$

$$L_{54} = (x_5 \frac{d}{dx_4} + x_4 \frac{d}{dx_5}) = (2x_4 \frac{d}{d(1 - r^2)} + \frac{1}{2}(1 - r^2)\frac{d}{dx_4})$$
$$= (-2x_4 \frac{d}{d(r^2)} + \frac{1}{2}(1 - r^2)\frac{d}{dx_4}), \quad \text{(III.4.c-9)}$$

$$L_{6j} = -(x_6 \frac{d}{dx_j} + x_j \frac{d}{dx_6}) = -(2x_j \frac{d}{d(1 + r^2)} + \frac{1}{2}(1 + r^2)\frac{d}{dx_j})$$
$$= -(2x_j \frac{d}{d(r^2)} + \frac{1}{2}(1 + r^2)\frac{d}{dx_j}), \quad \text{(III.4.c-10)}$$

$$L_{64} = i(x_6 \frac{d}{dx_4} - x_4 \frac{d}{dx_6}) = -i(2x_4 \frac{d}{d(1 + r^2)} - \frac{1}{2}(1 + r^2)\frac{d}{dx_4})$$
$$= -i(2x_4 \frac{d}{d(r^2)} - \frac{1}{2}(1 + r^2)\frac{d}{dx_4}), \quad \text{(III.4.c-11)}$$

$$L_{56} = (x_5 \frac{d}{dx_6} + x_6 \frac{d}{dx_5})$$
$$= ((1 - r^2)\frac{d}{d(1 + r^2)} + (1 + r^2)\frac{d}{d(1 - r^2)}) = -2r^2 \frac{d}{d(r^2)}. \quad \text{(III.4.c-12)}$$

Conformal generators (sec. III.4.a.iii, p. 138)

$$M_{\mu\nu} = L_{\mu\nu}, \quad \mu, \nu = 1, \ldots, 4, \tag{III.4.c-13}$$

$$P_\mu = id_\mu = (L_{5\mu} + L_{6\mu}), \tag{III.4.c-14}$$

$$D = -ix_\nu d_\nu = L_{65}, \tag{III.4.c-15}$$

$$K_\mu = 2i(2x_\mu x_\nu d_\nu - x^2 d_\mu) = (L_{5\mu} - L_{6\mu}) = 4ix_\mu \frac{d}{d(r^2)} - ir^2 \frac{d}{dx_\mu},$$
$$\tag{III.4.c-16}$$

obey the commutation relations of the conformal group and of SO(4,2).

To show this is the proper definition of variables we have to relate generators expressed in terms of them. How do these act on the variables? Internal parts are not considered. First the action has to be defined, with only that on x_ν being necessary, especially for

$$x_i \frac{d}{dx_5} = 2x_i \frac{d}{d(1-r^2)} = -2x_i \frac{d}{dr^2}, \tag{III.4.c-17}$$

$$x_i \frac{d}{dx_6} = 2x_i \frac{d}{d(1+r^2)} = 2x_i \frac{d}{dr^2}, \tag{III.4.c-18}$$

and

$$2x_i \frac{dx_\nu}{dr^2} = \frac{x_i}{x_\nu}. \tag{III.4.c-19}$$

What does this mean? These variables are regarded as matrices, so are invertible (eq. II.5.a–7, p. 99). The definition of the inverse is

$$\frac{1}{x_\nu} = \frac{x_\nu}{r^2}, \tag{III.4.c-20}$$

with the inverse of a component of the vector equal to the component divided by the magnitude of the vector. With this definition of the inverse the product of a component and its inverse is not unity, except if there is only one component. For more than one each is divided by the same value, and it is the sum of the products of the components with their inverses that equals 1. This results in

$$L_{6j}x_\nu = -(x_\nu \frac{x_j}{r^2} + \frac{1}{2}(1+r^2)\delta_{j\nu}), \tag{III.4.c-21}$$

$$L_{56}x_\nu = x_\nu. \tag{III.4.c-22}$$

Thus we get

$$L_{5,6\mu}x_\nu = i(\frac{1}{2}(1 \mp r^2)\frac{dx_\nu}{dx_\mu} \pm 2x_\mu \frac{dx_\nu}{dr^2}) = i(\mp x_\mu \frac{x_\nu}{r^2} - \frac{1}{2}(1 \mp r^2)\delta_{\nu\mu}),$$
$$\tag{III.4.c-23}$$

$$L_{5j}x_\nu = i(x_5\frac{dx_\nu}{dx_j} - x_j\frac{dx_\nu}{dx_5}) = i(2x_j\frac{dx_\nu}{d(1-r^2)} - \frac{1}{2}(1-r^2)\delta_{\nu j}),$$

$$\text{(III.4.c-24)}$$

$$L_{54}x_\nu = (x_5\frac{dx_\nu}{dx_4} + x_4\frac{dx_\nu}{dx_5}) = (-x_4\frac{dx_\nu}{dr^2} + \frac{1}{2}(1-r^2)\frac{dx_\nu}{dx_4}),$$

$$\text{(III.4.c-25)}$$

$$(L_{5\mu} - L_{6\mu})x_\nu = i(2x_\mu\frac{x_\nu}{r^2} - r^2\delta_{\mu\nu}). \qquad \text{(III.4.c-26)}$$

Then

$$M_{\mu\nu}r^2 = 0, \qquad \text{(III.4.c-27)}$$

$$P_\mu r^2 = 2ix_\mu = (L_{5\mu} + L_{6\mu})r^2, \qquad \text{(III.4.c-28)}$$

$$K_\mu r^2 = (L_{5\mu} - L_{6\mu})r^2 = 2i(2x_\mu - r^2 x_\mu), \qquad \text{(III.4.c-29)}$$

$$Dr^2 = ix_\nu d_\nu r^2 = 2ir^2 = L_{65}r^2. \qquad \text{(III.4.c-30)}$$

Acting on coordinates these are

$$M_{\mu\nu}x_\rho = L_{\mu\nu}x_\rho, \quad \mu,\nu,\rho = 1,\dots 4, \qquad \text{(III.4.c-31)}$$

$$M_{\mu\nu}x_{5,6} = M_{\mu\nu}\frac{1 \pm r^2}{2} = 0, \qquad \text{(III.4.c-32)}$$

$$P_\mu x_\rho = i\delta_{\mu\rho} = (L_{5\mu} + L_{6\mu})x_\rho, \qquad \text{(III.4.c-33)}$$

$$P_\mu x_{5,6} = P_\mu\frac{1 \pm r^2}{2} = ix_\mu = (L_{5\mu} + L_{6\mu})x_{5,6}, \qquad \text{(III.4.c-34)}$$

$$Dx_\rho = -ix_\rho = L_{65}x_\rho, \qquad \text{(III.4.c-35)}$$

$$Dx_{5,6} = \pm ir^2 = L_{65}x_{5,6}, \qquad \text{(III.4.c-36)}$$

$$K_\mu x_\rho = i(2x_\mu x_\nu d_\nu - x^2 d_\mu)x_\rho = 2i(2x_\mu x_\rho - x^2\delta_{\mu\rho})$$
$$= 2i(2x_\mu x_\rho - r^2\delta_{\mu\rho}) = (L_{5\mu} - L_{6\mu})x_\rho, \qquad \text{(III.4.c-37)}$$

$$K_\mu x_{5,6} = \pm ix_\mu r^2 = (L_{5\mu} - L_{6\mu})x_{5,6}. \qquad \text{(III.4.c-38)}$$

Unlike Lorentz generators, $L_{5,6;\mu}$ act nonlinearly. This defines the action of all L operators. These SO(4,2) generators are related to those of the conformal group, correlating the six and four variables of the groups, using this definition of their action.

III.4.d   Why the spaces are related as they are

The algebra of the conformal group, although acting in 3+1 dimensions, is isomorphic to the so(4,2) algebra. How are the geometries of these spaces related, why, and what justifies our choice of x_5 and x_6?

Essentially the four-dimensional space is taken as a subspace of that of six dimensions, but an unusual one, using a generalization of the projective plane. To explain this we first consider simple examples, for which the method does not fully work. But these introduce the ideas and terminology.

III.4.d.i   *The simple example of the projective line*

To clarify the meaning of embedding a space in a larger space, in the way that is done here, consider a plane, covered with radial lines from the origin. Each line, that is ray, is considered as a single point of the smaller space, which is thus one-dimensional. We wish to find functions on the plane that are functions over the subspace. They can depend only on the angle the ray makes with an arbitrary x axis, that is the ratio $\frac{x}{y}$. What functions are these?

Taking function $f(x,y)$, we divide by y, giving $y^a f(\frac{x}{y})$. The functions cannot have negative powers of y for that gives singularities violating rotational invariance, and also the choice of the y axis is arbitrary. The only functions we can consider are homogeneous ones, sums of terms $x^p y^q$, $k = p + q$, with the same power of k, and then choose as our functions $F(\frac{x}{y}) = f(\frac{x}{y})$.

Such a function is

$$f(x,y) = x^3 + x^2 y + y^2 x + y^3. \qquad (\text{III.4.d.i--1})$$

Factoring gives

$$f(\frac{x}{y}) = y^3 (\frac{x^3}{y^3} + \frac{x^2}{y^2} + \frac{x}{y} + 1), \qquad (\text{III.4.d.i--2})$$

and

$$F(\frac{x}{y}) = (\frac{x^3}{y^3} + \frac{x^2}{y^2} + \frac{x}{y} + 1), \qquad (\text{III.4.d.i--3})$$

a function only of the ratio $\frac{x}{y}$. However this does not extend to a function over the entire plane, that is in two variables. Thus we have to look at larger spaces.

III.4.d.ii   *The plane as an example*

Next consider the xy plane, in which we draw a circle about the origin, of radius r, for a definite-metric space, or a hyperbola for an indefinite-metric one (eq. II.2.c.-1, p. 74), and assign point

$$z = \frac{1 - r^2}{2},$$ (III.4.d.ii-1)

to the curve. From each point on it we draw a ray to this z, producing a cone. Transformations are taken to be in this manifold of rays, that is the space for which each ray is a point. The symmetry group of the curve is SO(2), or SO(1,1) for the two metrics. For each such curve we obtain a different set of rays, thus a different z.

This is shown in {cn := cone([0,0,-1],1,1,color=black): cnr := rotate(cn, 0, Pi, 0): display (cnr, axes=normal); cnl := cone([0,0,-2],2,2, color = black, style=point): cnlr := rotate(cnl,0,Pi,0): display (cnr,cnlr,axes = normal);},

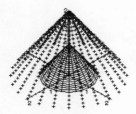

Figure III.4.d.ii-1: Cones, with each line a point in the projective space,

where we have two cones, one drawn with solid lines, the other larger one, with points. Each of these lines is a point in the projective space and each maps to a point on the z axis.

Going from one curve to the other is a translation, so the transformation group (which is not a symmetry group — emphasizing that transformation groups need not be symmetry groups) of this space of rays is SO(2) $\times T_1$, or SO(1,1) $\times T_1$, the direct product of the relevant rotation group and T_1, the translation group in one dimension, z. The generators are

$$J_{xy} = i(x\frac{d}{dy} - y\frac{d}{dx}), \quad \text{and} \quad P_z = id_z.$$ (III.4.d.ii-2)

This space also is insufficient but it introduces the terminology, and we now turn to the embedding of the conformal group.

III.4.d.iii *Embedding of R(3,1) inside R(4,2)*

The space in which the conformal group acts is $R(3,1)$, the (real) 3+1-dimensional Minkowski space, while the so(4,2) algebra to which the conformal algebra is isomorphic is defined over $R(4,2)$. We wish to embed our 3+1-space in the latter.

The radius of a sphere in $R(3)$ is

$$r^2 = x^2 + y^2 + z^2, \qquad \text{(III.4.d.iii-1)}$$

and we extend the $(x, y, z, t = x_4)$ coordinates to

$$(x_1, x_2, x_3, x_4, x_5, x_6) = (x, y, z, t, \frac{1 - r^2 + t^2}{2}, \frac{1 + r^2 - t^2}{2}).$$
$$\text{(III.4.d.iii-2)}$$

This gives

$$x_1^2 + x_2^2 + x_3^2 + x_5^2 = x_4^2 + x_6^2, \qquad \text{(III.4.d.iii-3)}$$

$$x_5 + x_6 = 1. \qquad \text{(III.4.d.iii-4)}$$

It is the latter equation (for a plane) that will allow us to define homogeneous coordinates, the reason these coordinates are defined as they are.

These six coordinates define a cone (a light cone, called here the nullcone) in the six-dimensional space, and we identify all points on each ray on this cone with a point in a smaller space, which therefore has dimension five. The intersection of this set of rays on the cone with the plane has dimension 3+1, and is taken to be the space in which we exist.

The coordinates are defined as

$$x = \frac{x_1}{(x_5 + x_6)}, \quad y = \frac{x_2}{(x_5 + x_6)}, \dots, \quad t = \frac{x_4}{(x_5 + x_6)}. \qquad \text{(III.4.d.iii-5)}$$

Clearly for our space this division has no effect. However it allows us to find functions over the six-dimensional space that can be restricted to the space of rays on the cone. These must be homogeneous in the 3+1 coordinates and are

$$F(x_\mu) = (x_5 + x_6)^a f(\frac{x_1}{x_5 + x_6}, \frac{x_2}{x_5 + x_6}, \frac{x_3}{x_5 + x_6}, \frac{x_4}{x_5 + x_6}),$$
$$\text{(III.4.d.iii-6)}$$

where the degree of homogeneity a, is the "conformal weight". In the plane

$$F(x_\mu)_p = f(x_1, x_2, x_3, x_4) = f(x, y, z, t), \qquad \text{(III.4.d.iii-7)}$$

so are functions over 3+1 space, but extend to functions over the entire six-dimensional space. This is then the embedding that we wish, and defines the functions over the six-dimensional space in terms of arbitrary ones in 3+1 space.

Group SO(4,2) leaves invariant the nullcone (as SO(3,1) leaves invariant the light cone) and the degree of homogeneity, since generators are of the form $x\frac{d}{dy} - \ldots$. Thus there is a unique function $g(x, y, z, t)$ such that

$$h(x_1, \ldots, x_6) = L_{ij}[(x_5 + x_6)^a f(x, y, z, t)]$$
$$= (x_5 + x_6)^a g(x, y, z, t) + g_o(x_1, \ldots, x_6), \quad\text{(III.4.d.iii–8)}$$

with L_{ij} an algebra operator and $g_o(x_1, \ldots, x_6)$ vanishing on the null-cone to preserve the conformal weight (on the cone, but not necessarily elsewhere).

Functions on the light cone (and similarly the nullcone) differ from others in being functions of fewer variables. Statefunctions of massive objects are (in principle) nonzero everywhere within the light cone, thus are functions of four variables. But the electromagnetic potential is defined, nonzero, only on the light cone, so this statefunction depends on three variables [Mirman (1995c), sec. 2.3.1, p. 15]. This leads to gauge transformations [Mirman (1995c), sec. 3.4, p. 43]. Lorentz transformations, and similarly those of the SU(2,2) and SO(4,2) groups, preserve the metrics, and cannot change the number of variables in the functions. The cones are invariant. Functions that are defined on these cones remain on them under transformations of the relevant group.

For example the photon, with coordinates,

$$x, \ y, \ \eta = z - t, \ \zeta = z + t, \quad\text{(III.4.d.iii–9)}$$

has (unnormalized) statefunctions, for photons moving in the two directions,

$$|1) = exp(ip_x x + p_y y + p_\eta \eta), \quad\text{(III.4.d.iii–10)}$$
$$|2) = exp(ip_x x + p_y y + p_\zeta \zeta), \quad\text{(III.4.d.iii–11)}$$

and each is a function of only three variables, not four as are statefunctions of massive particles. To mix η and ζ, as is required for the functions to depend on four variables, it is necessary to go off the light cone. Proper Lorentz transformations (not including inversions, which also do not change the number of variables) do not mix these states, so leave them both as functions of three variables, the same three. Nor are η and ζ mixed by transformations of the complete Lorentz group, as can be seen from their explicit realization [Mirman (1995c), sec. 2.3.1, p. 15]. For a massive object, not on the light cone, the four variables are mixed.

However plane $(x_5 + x_6)$ is not invariant. Functions cannot be defined only on it. Homogeneous functions on the intersection of the plane and the nullcone (which are functions of 3+1 variables) when transformed remain on the nullcone, but not on the plane. They are functions of five variables, and we are considering them only for particular sets of values, those on the plane. When transformed, functions for this range of variables remain functions of five variables, these no longer all referring to the plane. We now identify the five variables with the four of

Minkowski space, as we have done previously. Thus we are considering the transformed function again only for this set of variables, that is the 3+1-dimensional space. This gives a transformation of the four variables, so a transformation of the function defined over 3+1 space. We have related the action of nonlinear operators to those of linear ones.

III.4.d.iv   *The meaning of the relationship between variables*

Physical functions, basis states of the conformal group, statefunctions of physical objects, depend on four variables. How can we consider six? What does this procedure actually do?

We define two functions of the four spacetime variables, x_5 and x_6. Functions of the six coordinates cannot be physical functions but by restricting those of the six-dimensional space to ones with specific properties, and considering them only on a subspace, we obtain functions over our 3+1-dimensional space. There is a one-to-one relationship between conformal group basis states of the type of representations that we are studying, and the specified set of functions in the six-dimensional space, when restricted to the subspace.

Of course there is no six-dimensional space. It is purely a different mathematical way of giving functions over our 3+1-dimensional one, one that is useful for calculational purposes.

Why are x_5 and x_6 defined as they are? Clearly we want $x_5 + x_6 = 1$, and as we do not want to use them to pick out a direction in three-space, they have to be functions of r^2. Also they should be single-valued. The definitions are the simplest ones meeting these requirements. This does not imply that more complicated ones might not be of greater use in special situations, although we do not suggest that other choices are possible. Perhaps general proofs can be given limiting the set of definitions that work. But this choice is sufficient for the present purposes, and there is no need to consider others here.

The algebras of the conformal group, and of SO(4,2) on this space, are isomorphic. The four and the six variables are unique functions of each other. Thus when we perform an operation on the six-dimensional space (at least using the algebra), we also perform an equivalent one of the conformal group (on the functions in which the four variables are explicit), and one set of variables can be expressed in terms of the other, either or both before or after the transformations. These transformations must then give the same variables no matter how they are realized, in terms of one set, or in terms of the other. Functions of the variables are transformed the same way, no matter which set of variables they are written in terms of, and transformed functions of the four variables must be the same whether transformations were done using these variables, or using the six which are then expressed in terms of the four they are functions of.

Transformations of the six-dimensional space give functions that are not in the subspace. But they are equivalent to ones of the conformal group on the 3+1-dimensional space, though the action of the latter is much more complicated. However when we take the function resulting from a transformation in six-dimensional space, and replace the two variables by their expressions in terms of the variables of 3+1-dimensional space, we obtain the functions that we would have if we had just used a conformal transformation in the four variables. But the method of using a six-dimensional space in an intermediate step is simpler, since it uses linear transformations, not nonlinear ones.

III.4.d.v   *The relationship between spaces of 4 and 6 dimensions*

To find the relationship between operators of the two spaces, of 4 and of 6 real dimensions, we use the definition of the momentum operator, which produces translations, and defines the four-dimensional space,

$$P_\mu = (J_{5\mu} + J_{6\mu}).$$ (III.4.d.v–1)

Thus

$$exp[-i\theta(J_{5\mu} + J_{6\mu})] = exp(iP_\mu x_\mu) \Rightarrow exp(ip_\mu x_\mu).$$ (III.4.d.v–2)

With magnitude of the vector whose components are p_μ determining (the irrelevant scale) this can be taken as the definition of x_μ in terms of $z_1, ..., z_5, z_6$. It gives

$$p_\mu x_\mu \Rightarrow z_\mu cos\theta + z_5 sin\theta - z_4 cosh\theta - z_6 sinh\theta.$$ (III.4.d.v–3)

There are two arbitrary constants, say z_5 and z_6, since the space has two additional dimensions. The formulas are similar for x_1, x_2 and x_3, while for x_4, $5 \Leftrightarrow -6$.

III.4.d.vi   *The action on momentum eigenfunctions*

Representations of the conformal, so the Poincaré, groups that we are considering are those with the momentum operators diagonal. What is the effect of dilatation and transversion operators on their eigenfunctions? This also illustrates the discussions of embedding and of homogeneous functions.

Eigenfunctions of momentum are

$$|x, t\rangle = exp\{i\frac{x_1 p_1 + x_2 p_2 + x_3 p_3 - E x_4}{(x_5 + x_6)}\};$$ (III.4.d.vi–1)

these are the usual eigenfunctions on plane $(x_5 + x_6) = 1$, but extend to the entire six-dimensional space. We have to consider the action of

$$K_k = i(2x_k x_\nu d_\nu - x^2 d_k) = (J_{5k} - J_{6k})$$ (III.4.d.vi–2)

on them.

Derivatives are with respect the coordinates of the 3+1-dimensional space, so do not act on the denominator in the exponential. Thus

$$K_k |x, t) = i\frac{2x_k x_\nu p_\nu - x^2 p_k}{x_5 + x_6}|x, t).$$

These of course are not eigenfunctions of momentum operators.

The finite action is a rotation (and products), so for

$$x_1 = x_1' cos\theta + x_5' sin\theta, \quad x_5 = x_5' cos\theta - x_1' sin\theta, \qquad \text{(III.4.d.vi-3)}$$

the transformed function is

$$|x', t') = exp(i\{\frac{(x_1'cos\theta + x_5'sin\theta)p_1 + x_2 p_2 + x_3 p_3 - Ex_4}{(x_6 + x_5'cos\theta - x_1'sin\theta)}\})$$

$$= exp(i\{\frac{(x_1'cos\theta + \frac{1}{2}(1 - r'^2)sin\theta)p_1 + x_2 p_2 + x_3 p_3 - Ex_4}{(\frac{1}{2}(1 + r'^2) + (\frac{1}{2}(1 - r'^2)cos\theta - x_1'sin\theta)}\})$$

$$= exp(i\{\frac{x_1'cos\theta + \frac{1}{2}(1 - r'^2)sin\theta)p_1 + x_2 p_2 + x_3 p_3 - Ex_4}{\frac{1}{2}(1 + cos\theta) + \frac{1}{2}r'^2(1 - cos\theta) - x_1'sin\theta}\}).$$

$$\text{(III.4.d.vi-4)}$$

For rotations between the 1,2,3 and 6 axes, or between 4 and 5, the expressions are the same, except that $cos\theta \Rightarrow cosh\theta, sin\theta \Rightarrow sinh\theta$.

III.4.d.vii *Using these transformations to determine group representations*

In principle, how are these transformed states used to find group representations? First we summarize more familiar cases.

For the rotation group a transformed state is expanded in terms of basis states, spherical harmonics, functions of the transformed coordinates. And (by orthogonality) all states in an expansion have the same value of representation label l. Also there are a limited number of states in the expansion, given by $-l \le m \le l$, with m having integral spacing. This is the model for compact semisimple groups.

The Lorentz group, a model for noncompact, semisimple groups, allows a similar method (appendix A, p. 223). But the number of states is infinite for unitary representations. And the restrictions on state labels are different.

The inhomogeneous Poincaré group is somewhat distinct. It acts on two spaces simultaneously, an internal one with basis vectors that are basis states of the Lorentz subgroup, these multiplied by functions of space, exponentials for the representations considered here. The Lorentz part of the representations is handled as usual, with the second Poincaré invariant giving its representations. State labels of the space part are the momentum components, and these are subject to the requirement of the first Poincaré invariant: the sum of their squares equals the mass squared. Lorentz subgroup transformations mix momentum components, translations change phases of exponentials. Each state then is an exponential, times an internal state, given by a set of momentum components, and a phase. A transformed state is labeled by transformed momentum components and phase.

This is for free massive objects. Representations for massless objects, and with interactions, are more subtle and interesting [Mirman (1995c)]. In principle, the method is the same.

Conformal representations can be studied similarly: transforming states (here only one of the transformations is shown) and expanding them in terms of the basis states expressed in the transformed coordinates, the momentum eigenstates (which presumably can be done, although the differences between these transformed states and those of more familiar groups is so great this must be shown). Such expansions give conditions on the states in them, including their range, and matrix elements. Each set of states forms a representation.

This then is the procedure in outline. Clearly carrying it out, even for a few cases, would be a major calculational task, one too large to be even started here. It is likely however to produce information of some value, certainly mathematically, and probably physically.

III.4.e   Diagonalizing the operators

The usual familiar representations (which thus allow study of those of the conformal group) of SO(4,2), with operators $L_{\mu\nu}$, have L_{12}, L_{34} and L_{56} diagonal. For representations based on the Poincaré group they are not, it is the P's that are diagonal; these are sums of SO(4,2) generators, one compact, one noncompact (sec. III.4.a.iv, p. 139) — of a rotation and a boost. Thus to find the relevant representations, those that are representations of the Poincaré subgroup with diagonal momentum operators, from the known representations we have to diagonalize these sums. Basis states of the required representations are therefore sums of those of the familiar ones of SO(4,2) (sec. III.4.a.iii, p. 138), ones in which diagonal operators are either a rotation or a boost, but not a sum of these. It is the coefficients in such sums that are needed.

III.4.e.i   *Representations of* SO(2,1) *can be used*

The operators that must be diagonalized, the P's, are analogous to the N's [Mirman (1995c), sec. 2.3.1, p. 15]. These N's are sums of rotation and boost generators of the SO(2,1) Lie algebra, that of the Lorentz group in three-dimensional space, which is the algebra of SU(1,1), so this can be used to study them without considering the full SO(4,2) algebra. First we summarize the procedure, which is standard.

Matrix elements of SO(2,1) generators, for representations in which the L_{12} are diagonal, are known [Maekawa (1979a,b,1980)], say by continuation from those of SO(3), giving the matrix elements of the sums, the P's. An eigenvector of P is a sum over states of a representation of SO(2,1), with coefficients that we have to determine. We know the action of each term in the sum making up P on each of these states, thus of P. To get eigenstates of P we set the state obtained by the action of P on an eigenstate (with unknown coefficients) equal to the state, times a constant, the eigenvalue, and then use orthogonality to obtain the coefficients of the SO(2,1) eigenstates in the sum giving the eigenstate of P. With these we then have the eigenstates.

III.4.e.ii *Reviewing the diagonalization of an angular momentum operator*

Since diagonalizing momentum operators does not seem to be trivial (because they are sums of operators of different types), it is useful to review methods of diagonalization for familiar cases. As a simple example we consider SO(3) and diagonalize L_x starting from a basis in which L_z is diagonal, the spherical harmonics. Using the expression of L_x in terms of θ, ϕ [Varshalovich, Moskalev and Khersonskii (1988), p. 41], we carry out the differentiations [Varshalovich, Moskalev and Khersonskii (1988), p. 146], and use recursion relations [Varshalovich, Moskalev and Khersonskii (1988), p. 145], to get the action of L_x on $Y_{lm}(\theta, \phi)$, which is a sum of terms of the form $Y_{lm'} exp(\pm i\phi)$, all of the same l value. Thus we write the eigenvector of L_x as a sum over Y_{lk}, let L_x act on each term and then set the resultant sum equal to the original eigenvector, times a constant. Using the orthogonality relations [Mirman (1995a), sec. VII.5, p. 193] we obtain the coefficients in the sum that is the eigenvector, and the eigenvalue. This can be done for each l, for which there are $2l + 1$ eigenvectors.

Another way of doing this is to rotate axes so the x axis replaces the z axis. Since we know the transformation of the spherical harmonics under rotations [Varshalovich, Moskalev and Khersonskii (1988), p. 141] we then have the sums of spherical harmonics that are eigenvectors of L_x. The rotation gives

$$Y_{lm'}(\theta', \phi') = \sum_m Y_{lm}(\theta, f) D_{mm'l}(\alpha, \beta, \gamma), \qquad (III.4.e.ii-1)$$

where α, β, γ, are the Euler angles [Mirman (1995a), sec. X.4.a.ii, p. 278; Varshalovich, Moskalev and Khersonskii (1988), p. 22]; D is the Wigner D-function [Varshalovich, Moskalev and Khersonskii (1988), p. 72].

For the Euler angles, γ is the angle of rotation in the xz plane. For a boost y is imaginary, so in the Wigner D-function,

$$cos(\frac{y}{2}) \Rightarrow cosh(\frac{y}{2}), \quad sin(\frac{y}{2}) \Rightarrow sinh(\frac{y}{2}). \tag{III.4.e.ii-2}$$

This gives the transformation that diagonalizes the boost generator. Eigenstates are sums of L_z-diagonal states (of the Lorentz group) given by this procedure.

III.4.e.iii *Diagonalizing the P operators*

Consider a single P_μ (equivalent to N_μ). This depends on three variables (unlike familiar cases in which generators depend on two) so the L operators of SO(4,2) going with a single P belong to the 2+1-dimensional Lorentz group. This is true for all four P's, with different Lorentz groups, but these share two coordinates (x_5 and x_6). A P_μ is then (eq. III.4.a.iii-3, p. 138)

$$P_\mu = (L_{5\mu} + L_{6\mu}), \quad \mu = 1,\ldots,4. \tag{III.4.e.iii-1}$$

It acts on functions obtained from spherical harmonics by making one coordinate, the time, imaginary.

Note however that with

$$L_{ij} = -i(x_i\frac{d}{dx_j} - x_j\frac{d}{dx_i}), \quad i,j = 1,\ldots,6, \tag{III.4.e.iii-2}$$

we get for example

$$P_1 = (L_{51} + L_{61}) = -i(x_1\frac{d}{dx_5} - x_5\frac{d}{dx_1}) + i(x_1\frac{d}{dx_6} + x_6\frac{d}{dx_1}), \tag{III.4.e.iii-3}$$

$$P_2 = (L_{52} + L_{62}) = -i(x_2\frac{d}{dx_5} - x_5\frac{d}{dx_2}) + i(x_2\frac{d}{dx_6} + x_6\frac{d}{dx_2}), \tag{III.4.e.iii-4}$$

$$P_\mu P_\nu = 0. \tag{III.4.e.iii-5}$$

While the P's are required to commute, being inhomogeneous operators, here with this differential realization they obey a stronger condition, so the generators of the Poincaré subgroup cannot be realized this way. Perhaps this is not surprising since they must be realized as

$$P_\mu = i\frac{d}{dx_\mu}, \tag{III.4.e.iii-6}$$

and the two realizations, one homogeneous and one inhomogeneous, even with different variables, are not compatible.

Because the four P's commute, the eigenvectors found this way are compatible, they diagonalize all four P's together.

An eigenfunction of P is an infinite sum over these functions (the representation basis vectors of this Lorentz group), summed over all l and m values, with coefficients to be determined. The action of a P, written as a sum of the L's, on terms of such an infinite sum is thus known, so P gives another sum of these functions, with coefficients related to those of the sums making up the eigenfunctions. Setting two sums of a pair equal with a constant, the eigenvalue, and using the orthogonality conditions, gives the coefficients.

Another way of doing this is to take basis vectors of SO(2) and use the commutation relations of SO(2,1) — that group for which the P's are sums of its operators — to find the SO(2,1) basis states. Then each P is a sum of algebra operators of this group, a rotation and a boost, so its commutation relations with the other algebra operators are known. The sums then can be used to find the action of a P on a sum of these basis states, an eigenvector of the P, so this, with the orthogonality relations, gives its coefficients, and thus the eigenvectors of the P.

III.4.e.iv *Diagonalization of a specific P*

To illustrate this take P as a sum from the su(1,1) algebra of rotation L_{12} and boost L_{13} (sec. III.3.b, p. 133),

$$P = L_{12} + L_{13} = D + \frac{1}{2}(L_+ + L_-) = D + \frac{1}{2}(K + Q). \qquad \text{(III.4.e.iv-1)}$$

It is not a canonical generator of the algebra, but a sum of them, a reason its eigenstates are unusual. For state $|l, m\rangle$, these operators have the action

$$L_{12}|l, m\rangle = m|l, m\rangle, \qquad \text{(III.4.e.iv-2)}$$

$$L_{13}|l, m\rangle = M_+(l, m)|l, m + 1\rangle + M_-(l, m)|l, m - 1\rangle; \qquad \text{(III.4.e.iv-3)}$$

the M are their (known) matrix elements. Thus acting on an eigenstate P gives, with the c's the unknown coefficients, and Λ the eigenvalue,

$$P \sum_m c(l, m)|l, m\rangle = \Lambda \sum_m c(l, m)|l, m\rangle$$

$$= \sum_m c(l, m)\{m|l, m\rangle + M_+(l, m)|l, m + 1\rangle + M_-(l, m)|l, m - 1\rangle\}.$$

$$\text{(III.4.e.iv-4)}$$

Unitary representations of su(1,1) are infinite-dimensional, so these sums have an infinite number of terms. The states all belong to the same representation (which determines the matrix elements). We then get, by orthogonality,

$$\Lambda c(l, m) = mc(l, m) + c(l, m-1)M_+(l, m-1) + c(l, m+1)M_-(l, m+1).$$
$$\text{(III.4.e.iv-5)}$$

The M's are known. Also there are values m_m and m_x such that

$$M_-(l, m_m) = 0, \quad \text{or} \quad M_+(l, m_x) = 0. \tag{III.4.e.iv-6}$$

These give recursion relations which can be solved for the c's, determining a set of eigenvectors, one set for each real value of l.

The eigenvalues of P are continuous (not surprisingly since it involves a boost) there being no conditions to determine Λ.

For example, at the lowest state, $m_o = l$,

$$(\Lambda - m_o)c(l, m_o) = c(l, m_o + 1)M_-(l, m_o + 1). \tag{III.4.e.iv-7}$$

Normalizing so that $c(l, m_o) = 1$,

$$c(l, m_o + 1) = \frac{\Lambda - m_o}{M_-(l, m_o + 1)}, \tag{III.4.e.iv-8}$$

which can be written

$$\Lambda = M_-(l, m_o + 1)c(l, m_o + 1) + m_o. \tag{III.4.e.iv-9}$$

For the next state

$$\Lambda c(l, m_o + 1) = (m_o + 1)c(l, m_o + 1) + c(l, m_o)M_+(l, m_o)$$
$$+ c(l, m_o + 2)M_-(l, m_o + 2), \tag{III.4.e.iv-10}$$

$$\Lambda \frac{\Lambda - m_o}{M_-(l, m_o + 1)} - (m_o + 1)\frac{\Lambda - m_o}{M_-(l, m_o + 1)} - M_+(l, m_o)$$
$$= c(l, m_o + 2)M_-(l, m_o + 2), \tag{III.4.e.iv-11}$$

from which we find $c(l, m_o + 2)$. Also

$$c(l, -m_o - 1) = \frac{\Lambda + m_o}{M_+(l, -m_o - 1)}. \tag{III.4.e.iv-12}$$

As suggested by hermiticity we might want (but not necessarily get)

$$c(l, -m_o - 1) = c(l, m_o + 1), \tag{III.4.e.iv-13}$$

(otherwise the eigenstate would be two independent sums) so

$$\frac{\Lambda + m_o}{M_+(l, -m_o - 1)} = \frac{\Lambda - m_o}{M_-(l, m_o + 1)}, \tag{III.4.e.iv-14}$$

determining the value of Λ, if we require this.

III.4.e.v *Diagonalizing for the Pauli matrices*

As an illustrative example consider the defining representation of algebra su(2), which is so(3), with the (Pauli) matrices

$$\sigma_x = \begin{pmatrix} 0 & 1 \\ 1 & 0 \end{pmatrix}, \quad \sigma_y = \begin{pmatrix} 0 & i \\ -i & 0 \end{pmatrix}, \quad \text{(III.4.e.v-1)}$$

and we wish to diagonalize $\sigma_x + \sigma_y$. There are two eigenvalues, corresponding to the two of σ_z, $\pm\sqrt{2}$, with eigenvectors $(\pm\frac{1+i}{\sqrt{2}} \quad 1)$. These have complex entries; their conjugates are identical except that the sign of i is changed.

To obtain the diagonalization for so(2,1), set σ_x, σ_y to $i\sigma_x$, $i\sigma_y$, so that the matrix element for

$$\sigma_+ : \sqrt{l - m} \Rightarrow \sqrt{m - l}, \quad \text{(III.4.e.v-2)}$$

and that for

$$\sigma_- : \sqrt{l + m} \Rightarrow \sqrt{m + l}. \quad \text{(III.4.e.v-3)}$$

Thus, whereas for so(3), $m \leq l$, for this $l \leq m$ — these matrices are infinite-dimensional. It is infinite-dimensional matrices that have to be diagonalized, thus the actual details of the procedure may not be trivial, but perhaps the results might be interesting.

III.4.e.vi *Diagonalization of P on the Lorentz subgroup*

Group SO(2,1) can be realized as a subgroup of the Lorentz group whose matrix elements are known [Naimark (1964), p. 116]. We next consider the representations subduced [Mirman (1999), sec. VI.2, p. 281] from these Lorentz representations since we wish to consider the SO(2,1) group as a subgroup of a larger group, the conformal group and this provides an instructive example. Here then we realize the SO(2,1) group on a space larger than the 3 dimensional space of its definition, the 4-dimensional defining space of the Lorentz group (sec. A.7, p. 242).

We have, in Naimark's notation,

$$P = H_3 + F_+, \quad \text{(III.4.e.vi-1)}$$

a momentum operator as a sum of two operators of these representations (a rotation plus a boost). So, for a representation of the Lorentz group, thus of SO(2,1), with the representation labels omitted, ν the eigenvalue of H_3 and k the representation label of the SO(3) subgroup, we have, with the M's the matrix elements,

$$P \sum_{k\nu} c(k, \nu) | k, \nu) = \Lambda \sum_{k\nu} c(k, \nu) | k, \nu)$$

$$= \sum_{k\nu} c(k, \nu) \{\nu | k, \nu) + M_+^{k+1}(k, \nu) | k + 1, \nu + 1) + M_+^{k}(k, \nu) | k, \nu + 1)$$

$$+ M_+^{k-1}(k, \nu) | k - 1, \nu + 1) \}. \quad \text{(III.4.e.vi-2)}$$

This gives, by orthogonality,

$$\Lambda c(k,v) = vc(k,v)$$
$$+M_+^{k+1}(k-1,v-1)c(k-1,v-1) + M_+^k(k,v-1)c(k,v-1)$$
$$+M_+^{k-1}(k+1,v-1)c(k+1,v-1). \qquad \text{(III.4.e.vi-3)}$$

The matrix elements of F_+, the M's, are known, so we obtain a recursion relation to be solved for the c's and for Λ.

Taking one of the infinite-dimensional, unitary, representations we have the condition that at some $k = k_o = k'$ the M^{k+1} matrix elements are zero; the M_+ matrices are zero for $v = k$, and the M_- ones are zero for $v = -k$. We can take

$$c(k',k') = 1. \qquad \text{(III.4.e.vi-4)}$$

With the M's known, and k' depending on the representation, we have to find the c's, and Λ. There are three c's so there is some arbitrariness. For larger values of k, there are more c's, so more arbitrariness. There can therefore be many eigenstates which might perhaps have physical significance.

III.4.e.vii   *Relationship of the momentum eigenstate to the* SO(2,1) *states*

Starting for example with (eq. III.4.a.iii-3, p. 138)

$$P_1 = (L_{51} + L_{61}), \qquad \text{(III.4.e.vii-1)}$$

we wish to find $exp(ip_1x_1)$ in terms of the SO(2,1) basis states. State $exp(ip_1x_1)$ can be written $exp(-i(L_{51} + L_{61})x_1)$. Now we have

$$P \sum_m c(l,m)|l,m\rangle = \Lambda \sum_m c(l,m)|l,m\rangle, \qquad \text{(III.4.e.vii-2)}$$

where the c's are now knowable as we have just seen. Thus this sum is an eigenstate of P_1. But eigenstates of P_1 are $exp(ip_1x_1)$, so

$$exp(ip_1x_1) = \sum \sum_m c(l,m)|l,m\rangle, \qquad \text{(III.4.e.vii-3)}$$

where the first sum is over all eigenstates of P for this representation. Here l has any real value, so therefore does Λ, as does p. And $|l,m\rangle$ is a function of a single variable, as we have seen, so it can be made to correspond to x.

The representations are labeled p and l (sec. A.7, p. 242). To each representation of SO(2,1), given by l, there thus corresponds one of P, given by p, and conversely.

III.4.e.viii   *The relationship of the variables*

This procedure uses an expansion of basis vector $exp(ipx)$ in terms of basis states of SO(2,1). The former has one variable, x, the latter three. How are these related? Normalization of the states eliminates one variable, just as spherical harmonics depend on only two. Since $cos\theta \Rightarrow cosh\theta$, θ is a continuous variable ranging from $-\infty$ to ∞, as does x. There is also in the SO(2,1) basis states terms of the form $exp(i\phi)$, which just provide a phase. But the phase is absorbed into the coefficients in the eigenstate of P. Therefore the eigenstate depends on a single variable, matching thus $exp(ipx)$.

III.5   Implications and applications

These groups, the conformal group, and its smaller analogs, and their related groups, offer potential lessons, which undoubtedly the reader will more likely draw than will the author. But some are clear, and we suggest a few here.

One aspect appearing in even the simplest case (sec. III.3.a, p. 120) of some interest is that internal operators are multiplied by coordinates so their effect is different at different points. Also these internal operators change the space dependence of the basis vectors. For the Lorentz group it is the total angular momentum that is relevant, so (internal) spin and orbital angular momentum are coupled. But for the conformal group the relationship between internal and external parts of basis vectors is stronger. Unfortunately too little is known about the conformal group for the significance of this to be clear, yet it is certainly interesting, and might well be important.

III.5.a   The appearance of discreteness

How many eigenstates of P are there, for each representation? Each has a discrete, although infinite, number of basis states, so that should still be true no matter which operator is taken diagonal. Thus the number of independent eigenstates of P is also infinite, but discrete (for a representation, and different representations have different sets of eigenstates and eigenvalues). This is interesting in that the elementary-particle mass spectrum is discrete, and apparently infinite (and P can perhaps be a model of the Hamiltonian giving the rest mass). The spectrum is also given by integers (to a provocative approximation), as we see (appendix B, p. 246). Whether there are representations of the conformal group that give such spectra is not clear — obviously much study is needed. The present example is far too simple to serve as anything more than a suggestive model. But it illustrates the (possible) approaches, and what these might lead to.

There are also representations whose labels are discrete. Which of these types of representations, if any, might have analogs with physical applications in more realistic groups is not clear.

The spectrum of P may seem unusual. We might expect it to be continuous. However P is perhaps an unfamiliar generator, being a sum of a rotation and a boost. Also we are placing a perhaps unusual constraint on it. We found its spectrum not from regarding it as a member of the SO(2,1) algebra, but as a member of this subalgebra of a larger algebra, one with members having discrete spectra, and as a sum of generators of the larger algebra. Especially because of the form of the realization, with P a generator of a homogeneous algebra, but realized inhomogeneously, and with other generators, the K's, realized nonlinearly, we cannot find its properties from the subalgebra alone. In general, not only here, it is possible that representations of a group are different from those of a subgroup [Mirman (1995a), sec. X.6.b.ii, p. 290], with the latter not always what might be expected by subduction [Mirman (1999), sec. VI.2, p. 281]. We have to be cautious with beliefs that stem from more familiar groups when considering groups and realizations such as these.

III.5.b   Why generators are nonlinear

For Lie groups, we start from their algebra and exponentiate, giving transformations that are exponentials in the generators and transformation parameters. Yet conformal group transformations are not exponentials, but rather nonlinear ones, quotients. Why does the usual determination of the form of transformations fail for the K's? As we know (sec. III.3.a, p. 120) nonlinearity is also true for a realization of su(1,1). The reasons come not from the conformal group, but from the realization.

Nonlinearity appears because there are conditions on the transformation parameters. Operators are here functions of four real parameters, the spacetime coordinates. But the definition of the conformal group algebra uses a real six-dimensional space (so the group is SO(4,2)). Thus the operators should really be written in terms of six real parameters. The parameters we use, coordinates of 3+1 space, are then nonlinear functions of these, so the general method giving exponentials does not hold (sec. III.1.d.vi, p. 113).

Thus for example, we obtain relations between coordinates found by first defining (sec. III.4.c, p. 143)

$$r^2 = x_1^2 + \ldots + x_4^2, \tag{III.5.b-1}$$

where the x's are the coordinates of our four-dimensional space on which the Poincaré group acts, and then defining

$$x_5 = \frac{(1 - r^2)}{2}, \quad x_6 = \frac{(1 + r^2)}{2}, \tag{III.5.b-2}$$

so that

$$x_5 + x_6 = 1. \tag{III.5.b-3}$$

These are the coordinates of the space over which SO(4,2) acts and the parameters of the conformal group run over a subspace, a hypersurface, which places conditions on them leading to nonlinearities.

To get the Lie algebra, we apply the same method as usual, taking the transformations and expanding in a Taylor series, with the first term giving the algebra. When we do this for the nonlinear conformal transformations, we get nonlinear K's.

Another example is given by the expressions for the generators of the rotation group, using one or two variables rather than three real, or two complex, ones (sec. III.3.a, p. 120), and these contain nonlinear functions [Miller (1968), p. 45, 49].

For one dimension

$$K = ix^2 \frac{d}{dx}. \tag{III.5.b-4}$$

This is nonlinear. Why? Considered by itself it can be written

$$K = i\frac{d}{dy}, \quad \text{with } y = -\frac{1}{x}. \tag{III.5.b-5}$$

It is then linear (and Stone's theorem applies [Riesz and Sz.-Nagy (1990), p. 380]). Nonlinearity comes from choosing a variable that is a function of the one for which it is linear. However if we want it as a generator of a nonabelian algebra, the smallest one is that of su(2), or so(3), which requires two complex, or three real, variables. But here we use only one. Again it runs over a subspace, a line in the plane or volume, giving conditions on variables, reducing their number. This requires nonlinearity (for we know that we can realize the generators linearly using two complex or three real variables).

If we were able to realize generators linearly in fewer variables, then we would have more independent transformations in the subspace then it allows. There would be three sets of noncommuting rotations in two dimensions, which is not true. So the only way the generators can be realized over the subspace is nonlinearly.

In this nonlinear realization of the su(2) algebra the momentum operator, P, is realized as an inhomogeneous operator, rather than a homogeneous one as in the usual realizations of the algebra. How can a simple algebra contain an inhomogeneous generator? It is clear from this example that some generators can be inhomogeneous if others are nonlinear, and conversely. Inhomogeneity and nonlinearity are linked.

It is possible to pick a subalgebra of the (homogeneous, simple) conformal algebra which is the (inhomogeneous) Poincaré algebra because other generators are nonlinear.

The conformal group is a intrinsic property of geometry (sec. I.2, p. 9), and the Poincaré group is the transformation group of geometry [Mirman (2001), sec. I.3.b, p. 13]. That both are true (for the same geometry) results in some generators of the conformal group being (required to be) inhomogeneous, others nonlinear.

So the conformal group algebra is not a different algebra than that of so(4,2), but rather a different realization of it. Of course the group structure, being over a space of different dimension, is different. Details have to be investigated.

The necessity for nonlinearity can be visualized, not seriously of course, as due to the group transformations having to fit into a space too small for them, thus some have to be "crumpled up", becoming nonlinear.

III.5.c   Using what has been suggested by these groups

How do we use the relationships of these groups to find properties of the conformal group representations of the form that we wish? Since the invariants of SO(4,2) are known explicitly, or can be found, we can find the explicit expressions for those of these conformal representations. The eigenvalues of the invariants label the representations, providing information about them and the states.

In particular, we have discussed the expansion of $exp(ipx)$ in terms of the SU(1,1), or SO(2,1), states (sec. A.6, p. 240), and we know the invariants for this algebra, thus we can build on it to find those of the conformal group.

Actually carrying out the procedure explicitly is too involved to be done here; the likelihood is that it will lead to a great number of expressions, but the results might well be valuable. Among the questions whose answers could be suggestive are what are the eigenstates and eigenvalues of the invariants, and what are the condition on them? How do the facts that momentum is realized inhomogeneously and the K's nonlinearly affect the form and values of eigenstates and eigenvalues?

III.5.d   The unappreciated richness of group theory and possible implications

This indicates a little of the the richness of the representations of algebras (sec. A.10, p. 245), something usually not appreciated from the familiar, especially linear, representations alone. Once conditions like linearity are removed possibilities greatly increase. And as seen here there may be important physical applications that are not possible with more restricted forms.

That realizations of some generators are nonlinear raises interesting possibilities. Group representations play important roles in solving differential equations. In fact, their basis states are the solutions in many cases. And that solutions are basis vectors of representations of groups greatly restricts, so greatly aids in finding, them and their properties. But the equations of physics are (generally) nonlinear, although they might be modeled linearly. Might representation basis functions of nonlinear realizations be solutions of many of these nonlinear equations? That they are representation basis functions provides strong constraints, which helps us determine and understand them and their properties. It is not unreasonable to wonder whether study of groups with nonlinear realizations, like the conformal group, could lead to new and useful special functions, perhaps appearing not only as solutions of nonlinear differential equations. If this were the case it would open vast new fields for investigation. Because of this vastness, regrettably, all we can do now is suggest such possibilities.

These are important motivations for studying groups such as the conformal group, and these more general realizations and representations, which we can only start here, and for looking at experimental results, such as those for the masses (appendix B, p. 246), in a way that suggests relations between them and novel approaches to the mathematics.

Chapter IV

Conformal Field Theory

IV.1   Defining field theory

By a quantum field theory is usually meant a quantum mechanical system in which the number of objects can vary in time and space. In reality all fundamental physical systems, those not phenomenological [Mirman (2001), sec. I.4.c.v, p. 20], are described by field theories. Fields [Mirman (2001), sec. I.6.d, p. 32; sec. II.5.a, p. 83] are basis states of representations of the transformation group of space, and these are what statefunctions must be [Mirman (2001), sec. I.8.a, p. 44; sec. I.8.b, p. 48]. There really is no difference.

And, except for (apparently) only a few cases, isolated electrons, protons, neutrinos and photons, the number of objects in a state varies with time (the concept of number of "gravitons" is not well-defined [Mirman (1995c), sec. 11.2.2, p. 187]). Other single particles decay, and if there is more than one they interact: that is one absorbs or emits the other, or one emits an object that the other absorbs (correctly, as we discuss below, the statefunction of the system is a sum of those of its constituent objects plus terms that are products of statefunctions of these and of other objects).

It is sometimes said that quantum field theory differs from quantum mechanics in that a field theory has an infinite number of degrees of freedom, a quantum mechanical object has only a finite number. But the infinite number of degrees of freedom refers to massless objects, most familiarly the electromagnetic potential, photons. Certainly a field theory of the interaction between nucleons and pions does not describe a system with an infinite number of pions. Although there are certainly mathematical differences between these cases [Mirman (1995c), sec. 3.1, p. 32], it seems preferable at least at the fundamental level to consider all using a single formalism.

IV.1.a Interactions in quantum field theory

Fundamental equations (such as Dirac's) contain nonlinearities, the interaction terms, none contain potentials (scalars) or forces [Mirman (2001), sec. IV.1.a.iv, p. 145]. The term potential is used for two different entities, one — a fundamental physical object — is the electromagnetic potential, the statefunction of photons, the other its fourth component, which is a function not with respect to an arbitrary origin, but with respect to some fixed point, that is a physical object. Potentials, in this latter sense, such as in nonrelativistic Schrödinger's equations, are purely phenomenological. They are merely a means of modeling (simulating) interactions in a convenient way, but are not interactions — these are terms in the relevant equations nonlinear in the statefunctions so acting on a state they change the number of objects in it. The scalar electric potential [Mirman (1995c), sec. 7.2.2, p. 126], for example, is really due to the creation and annihilation of photons, whose number thus varies. And, as there is no such fundamental concept as potential, there is none that is force. Both are purely phenomenological. Force (which is not only a phenomenological concept, but a purely classical one as well) and potential never enter, except as expressions that are easier to work with then the proper ones, interactions of statefunctions.

Equations that are part of the foundations of physics are nonlinear. But these are difficult to deal with, actually impossible. How then can we express a field, that is its statefunction, to show that it obeys a nonlinear equation, and one connecting it to other fields, statefunctions of other objects? That is a central subject of this chapter.

IV.1.b Applications and extensions

Formalisms thus obtained can then be applied, for one to show why baryon number (so lepton number) must be conserved, why the proton cannot decay (sec. IV.4, p. 212).

And we are able to begin a study of the implications of the conformal group for quantum field theory, leading to conformal field theory, to find what hints and suggestions these give. How do conformal transformations, the conformal group, affect fields, how do they restrict them and laws governing them? As we have seen they imply the necessity of interactions, although it is not clear how these relate to known physical interactions, but they do offer clues (sec. I.7.d, p. 63). Likely much more will come from such investigations, which can only be glimpsed here. One purpose of this chapter, as for others in this book, is to raise such questions and possibilities, so to stimulate thinking, and hopefully greater understanding.

IV.2   How we mathematically describe fields

States of a field are basis functions of the transformation group (the geometrical transformation group [Mirman (1995b), sec. A.2, p. 178] which is perhaps a subgroup of a larger one), but of nonlinear representations. Thus to study a field we must express its states — basis states of nonlinear representations — in a way that includes variation of the number and types of objects they describe. Here the discussion is general, but to clarify its meaning we often use as examples specific elementary particles (usually just specific names). However there is nothing, except for such irrelevant names, to limit considerations to them.

This analysis is thus completely generic, describing states of a quantum field theory, and conditions on them. In specific applications it is necessary to be more explicit. But since we wish to consider quantum field theory as a framework [Mirman (2001), sec. I.4.a.iii, p. 16] such explicitness is unnecessary and might be distracting. Undoubtedly the reader will be able to transcribe the formalisms into the proper explicit ones for particular cases.

Field theories here are constructed (necessarily) using functions that are basis states of the transformation group of the geometry [Mirman (2001), sec. I.3.b, p. 13], the Poincaré group [Mirman (1995b), sec. 5.3.2, p. 103], and this is fundamental to the entire approach. However, as we know (chap. I, p. 1), there is a larger group, of which the Poincaré group is a subgroup, also closely related to the geometry, the conformal group. Thus it is interesting to study the requirement that field theory use its basis states, giving conformal field theory. Because this group, and its representations that seem relevant, are intricate and unfamiliar such study may lead to original and provocative results.

Certainly the requirement of invariance under the similitude subgroup gives conditions on the terms in equations (sec. I.2.a.iii, p. 11): they must all transform the same under it (colloquially, have the same units).

Since explorations of the group and its representations are complicated and novel, we merely just start to describe how the conformal group and quantum field theory can be related so hopefully creating (or at least suggesting) a foundation for studies of greater depth.

It is useful to begin by explaining concepts needed and their rationales. After that we can describe the formalism these concepts lead to, and how it is used. The formalism is always, necessarily, schematic but at first it is perhaps even more so.

IV.2.a   Where should interactions be inserted into theories?

Physical objects, being physical, must interact (or else we would not know of them, and if the objects of which we are composed did not interact, it would not matter that we were unaware of them). These objects, and their behavior, are described and governed by the Casimir invariants of the relevant groups [Mirman (1995b), sec. 3.4.4, p. 56; (2001), sec. II.5.a.ii, p. 84]. Thus these must contain interactions, nonlinearities. How can we put interactions in invariants?

Momentum operators (exponentiated) change statefunctions, functions of x_μ, to the corresponding functions of $x_\mu + a_\mu$; they are translations in space (generated by the three-momentum) and time (from the Hamiltonian). It is these that contain interactions; the Hamiltonian thus governs time behavior and this behavior is determined by the interactions that appear in the momentum operators of the object. Invariants of the Poincaré group are sums of products that include momentum operators, so must contain nonlinearities, and these nonlinear terms are given by those in the momentum operators.

IV.2.a.i   *The invariants for the Poincaré group*

We start with the Poincaré group, for which there are two invariants; these can be replaced by a single equation, Dirac's equation, for spin-$\frac{1}{2}$, but for that only. The invariants are P_μ^2, with eigenvalues

$$p_\mu^2 = m^2, \tag{IV.2.a.i-1}$$

which are, of course, constants, and W (sec. III.2.b.i, p. 119). Hence the values of the invariants are the mass (squared) given by the set of eigenvalues of the P's and that given by the total angular momentum (times the mass squared) in the rest frame. Into these invariant operators we have to insert interactions.

The argument for massless objects is similar, except for the little group [Mirman (1995c)].

IV.2.a.ii   *Interactions in the momenta give them in the invariants*

For a momentum generator we write

$$P_\nu = id_\nu + P_{\nu,int}, \tag{IV.2.a.ii-1}$$

so the invariants, acting on statefunction ψ, become, with m different from the value of the invariant for the free object (which really does not exist),

$$\sum P_\mu^2 \psi = -\sum d_\mu^2 \psi + \sum P_{\nu,int}^2 \psi + i \sum d_\mu (P_{\nu,int}\psi) + i \sum P_{\nu,int} d_\mu \psi$$
$$= m^2 \psi, \tag{IV.2.a.ii-2}$$

and

$$W\psi = \frac{1}{2}\sum M_{\mu\nu}M_{\mu\nu}m^2\psi - \sum M_{\mu\sigma}M_{\nu\sigma}(id_\nu + P_{\nu,int})(id_\mu + P_{\mu,int})\psi$$

$$= \frac{1}{2}\sum M_{\mu\nu}M_{\mu\nu}m^2\psi$$

$$-\sum M_{\mu\sigma}M_{\nu\sigma}(-d_\mu d_\nu\psi + P_{\nu,int}P_{\mu,int}\psi + P_{\nu,int}id_\mu\psi + id_\nu(P_{\mu,int}\psi))$$

$$= m^2 l(l + 1)\psi; \quad \text{(IV.2.a.ii-3)}$$

l is a pure number so W has the dimensions of m^2. The derivatives act on momenta because these contain statefunctions so are functions of space.

Eigenvalues of these, with interactions, are constants (under the group). This is a fundamental requirement on nonlinear representations, so on physical systems that they describe.

Of course finding these constants is, unlike the case for linear representations, a major (probably impossible) project, since these operators not only act on basis states (the solutions of the equations) but contain them. So we must at least start to explore how to deal with such nonlinear representations.

IV.2.a.iii   *Even stable objects have interactions*

Interactions appear, always, even for stable objects; the nucleon interacts with the pion, but cannot decay into states containing it, likewise the pion cannot decay (because of the strong interactions). Momentum operators and labeling operators contain interactions between these because the forms of operators is independent of the particular state acted on. If a state contains a nucleon and a pion they could interact, scatter, so momentum operators must include interaction terms that give such behavior. Equations governing objects are the same whether other particles are present or not, thus independent of the presence of other objects.

In general then a system is determined by the group (the Poincaré group, and supergroups of it perhaps like the conformal group, and ones involving internal variables such as isospin), and described by one of its irreducible representations [Mirman (2001), sec. I.8.b, p. 48]. Thus equations governing a system are the invariants of the group and the state labeling operators, these set equal to constants [Mirman (1995c), sec. 9.3.6, p. 164]. We usually consider representations in which momentum operators are diagonal. Interactions appear in these, and since other labeling operators are functions of them, nonlinearities of the momentum operators determine nonlinearities of the others.

IV.2.b   What are the states of a system?

To consider how transformation groups lead to descriptions of, and place requirements on, a system, we have to describe it, that is tell how to give its statefunctions [Mirman (2001), sec. II.5.a.ii, p. 84]. A statefunction is an eigenstate of the two Poincaré invariants (and others if there are more groups), and in addition we (here) wish to find statefunctions that are eigenstates of the four momentum operators. For a spin-$\frac{1}{2}$ particle these are the solutions of the Dirac equation, but in general one equation cannot substitute for the set [Mirman (1995b), sec. 6.3.2, p. 116]. Physical objects interact, and statefunctions thus must be of interacting objects. The problem is then to obtain expressions for such functions, say solutions of Dirac's equation with nonlinear terms, or generally solutions of the differential equations realizing the representation and state labeling operators, including the momenta.

The states that we use therefore are momentum eigenfunctions, but since we are considering conformal field theory these are eigenfunctions of the more general momentum operators (sec. I.3.d.iii, p. 32; sec. I.3.d.iv, p. 34), ones expressed in elliptic coordinates (sec. II.4, p. 96), the proper ones to use for accelerated objects, and the proper ones to use when considering the conformal group, so conformal field theory (although these really refer to two-dimensions; more general ones have to be developed (sec. I.5.g, p. 50)). Those for rectangular coordinates are a special case. We leave open the question whether these general eigenfunctions can be expressed in terms of the rectangular ones, and in particular what the significance of such expansions, and the coefficients therein, might be.

Because this opens the possibility of a vast field for exploration we limit here considerations of the effects of conformal field theory to those of the similitude subgroup, thus do not study the implications for physics of the requirement that these more general momentum eigenfunctions be the basis states on which the theory is constructed. Yet clearly using such general functions could well have important value.

IV.2.b.i   *What determines a system?*

The lists of particles, their masses, spins, internal quantum numbers, and of interaction terms, which particles interact with which, the coupling constants, and the way the spins and other quantum numbers appear, define the systems [Particle Data Group (2000)]. This definition is assumed known (although of course it is far from completely so) and that is what we start with. There is much structure in this data, some well-known, other aspects perhaps less so (appendix. B, p. 246), but since that is not understood we do not try to include it here.

The other part of the definition of the system is the group under which it transforms, (a supergroup of) the Poincaré group. Here we consider the conformal group, with algebra isomorphic to SO(4,2), this being fixed (here) we must define each system we study by giving its momentum operators.

IV.2.b.ii   *The vacuum state in the formalism*

This formalism requires a vacuum state, $|0)$. It is defined as that state on which all annihilation operators give zero, and has no physical implications (here) beyond that — there is no implication that this describes a physical system, or has any physical properties [Mirman (2001), sec. V.7, p. 251]. It is simply a symbol introduced to allow a convenient way of giving the formalism. Its only properties are that it has a conjugate vacuum state, $(0|$, that it is normalized (or normalizable),

$$(0|0) = 1, \qquad\qquad (IV.2.b.ii\text{-}1)$$

and that every annihilation operator, a, acting on it gives zero,

$$a|0) = 0, \qquad\qquad (IV.2.b.ii\text{-}2)$$

and likewise for the creation operators, a^*,

$$(0|a^* = 0. \qquad\qquad (IV.2.b.ii\text{-}3)$$

It is assumed that all states needed to give the group representations, and needed to properly describe physics, can be given by (functions of) creation and annihilation operators acting on this vacuum.

Also it is a fundamental assumption of this entire approach to the development of the representations that there is such a vacuum state, and that it is unique. This assumption is physically motivated. We can find, experimentally, states with zero electrons, zero pions, zero nucleons, and so forth, and also states with only one electron, or one proton, or one pion, and so on. Such states do exist experimentally, and no other states have ever been observed (except ones expandable in terms of these). Nor is it clear how they could be, and likely they cannot. There is no way of distinguishing a state of zero electrons from another with zero electrons (ignoring, of course, other objects that may be present). Nor is there a way of constructing mathematical theories with correct physical consequences that distinguish such states with zero objects. States with zero objects are physically and mathematically unique. And all states with one electron, for example, are identical if all parameters, like position, momentum and spin, are the same. Thus we need, and can use, only states that are identical if their particle numbers and parameters are. For massless objects, like photons, it might be difficult experimentally to distinguish between situations in which there are no photons, and ones in which photons are present but have extremely low energies (something not possible for massive objects). However that does not seem like an essential complication here, but more of an experimental problem.

Mathematically (as shown by, say, Haag's theorem) there are formalisms that have similarities to the present one, in which there are no unique vacuum states [Friedrichs (1953); Garding and Wightman (1954); Jauch and Rohrlich (1976), p. 503; Streater and Wightman (1964), p. 161; Wightman and Schweber (1955), p. 819]. This raises several possibilities. Mathematics often allows systems that have no relationship to physics, nor need they. Mathematics is much richer than physics (fortunately). It is not clear that field theories that do not have a unique vacuum allow representations of (the geometrical transformation) groups, and if not they would have no physical relevance (to the physical subjects studied here). Nor, since it is experimentally possible to find states with no objects, ones of the form considered here are clearly capable of describing such physical situations. It is also possible that these other forms of states do allow different ways of describing physical theories that can be in accord with experiment, but that are equivalent to formalisms considered here, or even inequivalent and that such different formalisms can lead to useful results. Or there may be aspects that the present approach cannot deal with, that may not have been noticed, but that other methods can (more successfully). However since this approach does seem to be in accord with our physical intuition and does seem capable of describing physics correctly (perhaps even fully) leading to fruitful consequences, and since mathematical structures are far richer than physical ones, it appears unnecessary — here — to consider other forms of field theories (if there are any).

IV.2.b.iii *The vacuum must be empty, so requires great care*

It is essential to emphasize that the vacuum does not have physical properties [Mirman (2001), sec. V.7, p. 251]. There are very strange beliefs about it [Glanz (1998); Quercigh and Rafelski (2000); Quigg (1999); von Baeyer (2000); Voss (1999); Weiss (2000)] such as that there are fluctuations in it, that it has energy, that it can be melted, that it consists of such things as spacetime foam, or that particles pop in and out of the vacuum. It has even been claimed that this has been experimentally verified [Kleppner and Jackiw (2000); Yam (1997)]! These beliefs are used to "explain" the triviality that two different equations, Dirac's equation without and with interactions, have different solution sets. This is exactly the same as saying that the reason the values of x that are solutions of the equations

$$5x = 7 \quad \text{and} \quad 23x = 34, \qquad\qquad \text{(IV.2.b.iii–1)}$$

are different is that particles pop in and out of the vacuum and that this changes the value.

It is a strange, and almost incomprehensible, aspect of pathopsychology that so many physicists (and journalists) have such an uncontrollable compulsion to make fools of themselves in public, as seen from these ideas about the vacuum and also (not only) the many examples mentioned before [Mirman (2001)]. What pops in and out of a vacuum is not particles, but such ideas — which unfortunately have too much energy.

IV.2.b.iv *Writing states with creation and annihilation operators*

States are group representation states, and these must be determined for particular cases being studied. Specifying them is a major part of setting up a theory. But first a general formalism — taking into account interactions — is needed. States without interactions will be written using creation operators, a^*, with the proper indices, acting on the vacuum. This correspondence between states and creation operators defines these operators (it being assumed that it is always possible to define a one-to-one correspondence). Each creation operator a_i^* has a conjugate a_i, an annihilation operator — defined by this; these have commutation relations

$$[a_i, a_j^*] = \delta_{ij},$$ (IV.2.b.iv-1)

and each commutes with all other operators. There is also a number operator (for each type of object)

$$N_i = a_i^* a_i.$$ (IV.2.b.iv-2)

The (nonphysical) free states are then

$$|i, j, \ldots) = a_i^* a_j^* \ldots |0),$$ (IV.2.b.iv-3)

for a state of several objects, these given by indices on the creation operators.

Questions have been raised about the hermiticity of objects such as creation and annihilation operators because

$$a|0) = 0, \quad \text{so} \quad a^* a|0) = 0, \quad \text{but} \quad aa^*|0) \neq 0.$$ (IV.2.b.iv-4)

However

$$(0|a^* a = 0,$$ (IV.2.b.iv-5)

and

$$(0|a^* a|0) = 0, \quad \text{while} \quad (0|aa^*|0) = 1.$$ (IV.2.b.iv-6)

So both expectation values are real. (However there could be subtle mathematical points [Vaccaro (1995)] that are beyond the scope of this discussion, but which might lead to interesting insight into group representations.)

IV.2.b.v *Physical creation and annihilation operators are unknown*

This form for states is not useful as we do not know the physical creation and annihilation operators, which we denote by A's, nor do we have any way of directly specifying their properties. It is not obvious that we can specify their commutation relations. A fundamental assumption here, but a much weaker one, is that these properties can be given for the "free" operators, with states

$$|\Delta, p_v, v, x) = exp(ip_v x_v) a^*_{\Delta v} |0), \qquad \text{(IV.2.b.v-1)}$$

where Δ is the symbol for the object (giving the relevant quantum numbers), p_v is the momentum, and v represents any other variables. Since free states are group representation basis states, of linear realizations, the commutation relations follow from the properties of group representations, except perhaps in very unusual cases, if there can be any. However physical operators go with nonlinear representations, much less is known about them, and they are more likely to have unexpected and unusual properties, so it is unwise to postulate such conditions as commutation relations for them.

Operators for physical particles, the A's, are unknown, including their commutation relations (such as at different points). Their commutation relations at different points, for example, do not follow in an obvious way from those at the same point. We do not know whether they form a complete set, or not, or whether they might give an overcomplete one (which is really true of the a's also, but these, being simpler, allow firmer assumptions). Also physical creation operators create states that are solutions of nonlinear equations, containing interactions. Much less is known about these than about solutions of linear equations, so they may not have properties that are the same as for the latter, and thus have unknown properties.

IV.2.b.vi *The Hilbert space for physical states*

Another reason that their properties are less obvious than those of operators for free states is that A's create and annihilate sums (including integrals) of products of states. That is we visualize the physical state as a sum of states, the free Δ, the state of the pair of particles $\Delta\pi$, of states $\Delta\pi\pi$, $\Delta\pi\pi\pi$, and so on. While this visualization may not be correct, it does warn us that we cannot assume properties of physical states as confidently as for free ones. Nevertheless it is necessary to have expressions for physical states, and we base it on this visualization. Thus this is a fundamental assumption of the present approach. And it does appear to be in accord with our intuition about what experiment shows. It seems difficult, or impossible, to interpret experiment differently.

The space over which the physical states are defined is then a sum of a Hilbert space of one particle, a Hilbert space (product) of two (perhaps different types of) particles, one of three, and so on, thus of an infinite number of spaces. There are well-known problems with such a space [Schweber (1962), p. 163], a reason we have to be careful of any assumptions about it.

But we need properties of states so must assume expressions for them — ones as general as possible. This visualization is used to obtain these.

IV.2.b.vii *States are eigenstates of the momentum operators*

We want eigenstates of the four momentum operators, and start with a state (schematically) of the form (ignoring normalization)

$$|\Delta_p, p_\nu, x) = exp(ip_\nu x_\nu)A_\Delta^*(p_\nu)|0), \qquad \text{(IV.2.b.vii–1)}$$

with Δ_p indicating the physical object, and $\nu = 1, \ldots, 4$. This is not a momentum eigenstate — momentum operators (including the Hamiltonian) have nonlinear terms which change objects in the state. We consider that, for example, Δ_p decays into a nucleon plus pions, so that a state with a single Δ_p is not invariant under time translation, thus is not a momentum (Hamiltonian) eigenstate, nor is it invariant under space translation, so is not an eigenstate of the three-momentum operators [Mirman (2001), sec. I.7.b.vii, p. 41]. Also the other operators (specifically transversion operators, K's) acting on these produce states that are not such eigenstates. Generally the free state is a wavepacket

$$|\Delta_p, x) = \int dp_\mu exp(ip_\nu x_\nu)f(p_\mu)A_\Delta^*(p_\mu)|0), \qquad \text{(IV.2.b.vii–2)}$$

and physical states have to be built using these. There are subscripts, which can be put in for specific cases, on creation and annihilation operators for spin components and other variables, like isospin (which presumably are not relevant here).

Creation operators $A_\Delta^*(p_\mu)$ create states of momentum p_μ and spin m, which gives both total spin (although this may be suppressed as fixed) and its component along some arbitrary axis. While the dependence on momentum is written explicitly, can the internal state in addition be a function of momentum? We assume (presumably without limiting generality) that if it is it can be expanded in a Fourier series, and then its momentum dependence is shown, and can be attached to the part of the state that has this already given. Therefore we can take the internal part of the state as independent of momentum, so write the operator as A_Δ^* (unless momentum dependence is not otherwise shown).

IV.2.b.viii *Wavepackets in momentum-diagonal representations*

A wavepacket is of course not a momentum eigenfunction; why should it be a state of a representation for which momenta are diagonal? It is a sum of momentum eigenfunctions, a sum of states labeled by their momentum values. To give a wavepacket we must give the coefficients in the sum — there is no label for a wavepacket that is an eigenvalue of an operator obtained from the algebra. Usually each momentum eigenstate in the wavepacket can be considered separately; this is the reason that the representation is considered to have momenta diagonal — the representation is built on momentum eigenstates, and physical states are sums of these. If we took other operators diagonal, such as those of angular momentum, then states would be labeled by angular momentum eigenvalues, and an expansion in terms of momentum eigenstates would contain an infinite number of terms, with (in principle) known coefficients. Terms in the sum could not be considered separately, for then we would not be using angular momentum eigenstates.

For the Poincaré group, and more general ones, it is not clear whether eigenstates of one set of operators can always be expanded in terms of those of another set. In some cases these types of states are so different [Mirman (1995c), sec. 2.3.2, p. 17] that their relationship is not obvious. Of course, the question whether the states are expandable in terms of each other is quite interesting, but (unfortunately) is not a topic we can consider here.

For a free particle (an unphysical one without interactions), the momentum eigenstates are known, and a general state, a wavepacket, is a sum of these. But for an actual physical object momentum eigenstates are sums over states of different particle number that are solutions of (in reality unsolvable) nonlinear equations. Thus we have a complete set of known solutions for free particles, so can consider only them. However states of physical ones are what we want but cannot find because we cannot solve the equations. But we can, and here try to, write formal expressions for them, using our knowledge of the free states.

IV.2.b.ix *The action of momentum operators*

How does momentum, expressed in terms of creation and annihilation operators, translate an object, that is (exponentiated) acting on a basis state that is a function of $x - 0$ give one that is a function of $x - x_o$? Consider state

$$|1) = A^*(p_\rho; \sum p_\rho^2 = m^2)|0), \qquad \text{(IV.2.b.ix-1)}$$

and the action of P_μ on it,

$$|1)' = P_\mu|1) = p_\mu A^*(p_\nu)|0), \qquad \text{(IV.2.b.ix-2)}$$

so that continuing,

$$exp(iP_\mu x_\mu)|1) = |1)' = (1 + x_\mu P_\mu + \frac{1}{2}x_\mu^2 P_\mu^2 + \ldots)A^*(p_v)|0)$$

$$= (1 + x_\mu p_\mu + \frac{1}{2}x_\mu^2 p_\mu^2 + \ldots)A^*(p_v)|0) \Rightarrow exp(ip_\mu x_\mu)A^*(p_v)|0).$$

$$\text{(IV.2.b.ix-3)}$$

Thus it is the presence of p_μ as a coefficient that leads to the state being translated.

But P_μ gives translation in the μ direction. However translation in one direction is related to — necessitates — translations in all (in a general coordinate system).

IV.2.b.x   *Why the Hamiltonian gives motion in space*

Why does the Hamiltonian H, the time-translation generator, in addition give motion in space? It is related to the three-momentum operators by the invariant,

$$H^2 - \sum P_i^2 = m^2. \qquad \text{(IV.2.b.x-1)}$$

Since H (exponentiated) moves the object in time, for this invariant acting on a statefunction to give a constant, three-momentum P, if it does not give zero on the statefunction (as it does in the rest frame), must move the object in space.

With m now the eigenvalue, rather than it times the unit operator, the statefunction is (schematically), with v the speed, and c, f, the proper coefficients,

$$|x,t) = \int d\omega dp c(p,\omega)exp(ipx + i\omega t) = \int dp f(p)exp(ip(x - vt)),$$

$$\text{(IV.2.b.x-2)}$$

so a displacement in time requires one in space; a surface of constant phase inclined to the time axis must also be inclined to a space axis.

If we realize momentum generators using creation and annihilation operators then, to show dependence on space, we must regard the action of a creation operator as giving the state

$$|1) = A_p^*|0) \Rightarrow exp(ipx)A_p^*|0); \qquad \text{(IV.2.b.x-3)}$$

therefore a Hamiltonian of the form

$$H = \sum \omega A_\omega^* A_\omega, \qquad \text{(IV.2.b.x-4)}$$

has the same effect as the derivative realization.

IV.2.b.xi *How spin and momentum can both be diagonal*

Creation operators act also on internal states; a particle in a momentum eigenstate has spin, so its statevector is a function over two spaces, real space, that in which we live, with the function being schematically of the form $exp(ipx)$, and spin space. Thus the creation operator gives a particle of spin s and z component s_z. Both spin and momentum are diagonal, which is possible as they act in different spaces — but if momentum is diagonal, orbital angular momentum cannot be, the state is then a sum over spherical harmonics. Because there are two spaces we thus (can) write states in this form (as is usual).

IV.2.b.xii *Action of group operators on interactions*

Interaction terms are products of statevectors, functions of space. How do the group operators act on them? For example, the term in momentum component P_μ for the interaction of the electromagnetic potential with an object is given by expressions like $A_\mu(x)\psi(x)$, where ψ is the statefunction of the charged object. This can be written in terms of creation operators for the electromagnetic potential, a_μ^*, and object, b_i^*, where i labels spin, as

$$H_v \sim \int dp\,dq\,C(p,q)a_\mu^* b_i^* exp(iqx)exp(ipx). \qquad \text{(IV.2.b.xii–1)}$$

There is a coefficient because states that appear in interaction terms are physical states, and their expansion depends on the particular solution.

Algebra operator $M_{\mu\nu}$, for example, is realized as a sum of three terms (here, but more if there are other fields),

$$M_{\mu\nu} = i(x_\mu d_\nu - x_\nu d_\mu) + i(A_\nu^* A_\mu + A_\mu^* A_\nu) + i(B_\mu^* B_\nu + B_\nu^* B_\mu), \qquad \text{(IV.2.b.xii–2)}$$

the first term (exponentiated) acting on x gives x' (that is it changes coordinate values in the basis vectors), the second changes the component of the electromagnetic potential to give the correct component for the momentum produced from the initial one by this operator, and the third likewise acts on the object's statefunction. Thus the group operators act on all functions, treating interaction terms as just products of functions. In this sense, all terms, whether in interactions or not, are treated equivalently.

IV.2.c What is the meaning of decay?

It is common to describe an object that decays using a complex Hamiltonian. Obviously this is purely phenomenological. The state of an object is a sum of eigenstates of that part of the Hamiltonian without the interaction causing the decay, with all states with which it interacts in the sum, so that the sum (of products of states) includes multiparticle states — for a decaying particle, the statefunction of the system is a sum of (the statefunctions for) the single-particle state, plus a sum over decay products. The eigenvalue of the total Hamiltonian for this state is of course real.

Taking the mass complex is merely a phenomenological way of representing this sum, and it should be understood as such otherwise some of the tools for studying properties of these states will be lost, such as inclusion of decay products, and is likely to cause confusion. And, of course, the approach, if taken seriously, would violate the fundamental principles of quantum mechanics and of group theory [Mirman (2001), sec. II.1.b.iv, p. 58].

IV.2.c.i *Describing systems with decay*

The correct description of the system, with state $|s)$, is (always schematically),

$$|s) = a(t)|p) + b(t)|decayproducts), \qquad (IV.2.c.i-1)$$

$$a(0) = 1, \quad b(0) = 0, \quad |a|^2 + |b|^2 = 1; \qquad (IV.2.c.i-2)$$

$|p)$ is the state of the decaying object, and $|decayproducts)$ the state giving the decay products (a sum of states in general, but written here as one), with the states orthogonal,

$$(p|decayproducts) = 0. \qquad (IV.2.c.i-3)$$

Coefficients a, b give the probability of finding the states, thus the probability of decay, as a function of time.

Experimentally, at any time, we see only one of the states in the sum. The probability that the particle did not decay is given by $(p|s)$, that it did by $(decayproducts|s)$. The state is a superposition only in the sense that it gives this probability; but if this system interacts with another the state appearing in the interaction is this superposition. Viewed this way it is exactly the same as a sum of statefunctions of a system that is not an eigenstate of some operator — it is a sum of eigenstates.

IV.2.c.ii *Motion of decay products*

For a particle that decays the decay products, whose statefunctions are wavepackets, move apart, so their coordinates (the centers of their wavepackets) become different. While all statefunctions are functions of the same variables, the coordinates of space, their wavepackets are (almost) confined to different regions.

Momentum operators consist of sums of parts, one for the initial particle, the others for each of the particles resulting from the decay. Every wavepacket is acted on by a different term in each sum making up a momentum operator. These cause the objects, that is the centers of their wavepackets, to diverge. It may be useful at times to measure positions of objects from the centers of individual wavepackets, thus using different coordinates for each. But they are all functions of the same set of coordinates, those of space — all exist in the same space [Mirman (2001), sec. III.3.d, p. 127]; it would be difficult for an object in one space to turn into several objects in different spaces.

IV.2.d   The general form of states

By a quantum field theory we imply a description of systems in which the number of particles change, in space and time. We need then a representation of a general state, one describing a totality consisting of a variable number of objects, and an eigenstate of the momentum operators (and of all other labeling operators). Basic ideas used to develop this have now been introduced and we can therefore proceed to the next small step.

To illustrate consider a simple system, an object, a fermion (say with spin-$\frac{1}{2}$) called Δ, with mass m_Δ for the free particle (a parameter used to facilitate labeling and discussion but with no physical meaning and whose value is thus irrelevant) and M_Δ for the physical (experimental) one; Δ interacts with N (a fermion) and π (a boson). Labels Δ, N and π are for statefunctions obeying equations without interactions, Δ_p, N_p and π_p label solutions of those with interactions, the physical objects. A Δ_p can decay into an N, with masses m_N and M_N, plus π's with masses m_π and M_π. Creation and annihilation operators labeled a_Δ, a_N, a_π, are for free particles; these give the corresponding states. Operators A_Δ, A_N, A_π are for physical objects, ones obeying equations with interactions. Momentum operators including the Hamiltonian, so Dirac's equation for the physical Δ, then include (here) these interactions with π and with N. Free equations include neither.

We limit the number of objects, here three, and the number of interactions, to prevent the notation from becoming overly complicated. But these arguments can be immediately extended to full generality by introducing more symbols, including indices. This presents the procedure and serves as a pattern for formalisms for specific cases (in principle; in reality there are a large number of different objects, possibly an infinite number, coupled to each one).

IV.2.d.i   *The expansion for nondecaying physical states*

Required state $|\Delta_p\rangle$ obeys a nonlinear equation, and one containing
statefunctions of different objects, so cannot be studied directly. We
start thus with ones obeying equations for free objects — without interactions — and indicate how statefunctions are written for stable particles.

The two simplest (known) objects (excluding neutrinos, photons and
"gravitons") are the physical proton, P_p, and physical electron; these are
stable and are taken as momentum eigenstates (or sums if wavepackets). Statefunctions for P_p while momentum eigenstates, are not the
same as for a free proton, P — they obey a different Dirac equation,
one with interactions (limited here to pions) rather than one without.
Thus we write the proton state, with integrals such as over momenta
and sums over variables such as spin suppressed, as (schematically)

$$|P_p\rangle = c_o(x,t)|P\rangle + \sum c_n(x,t)|P\rangle|\pi_p\rangle^n, \qquad \text{(IV.2.d.i-1)}$$

summing over the number of pions from one to infinity. There is a
similar sum for π_p. If there were no interactions only the first term
would appear. Interactions and initial conditions determine the c's.

In principle then this is substituted in Dirac's equation containing
the interaction between the proton and pion, with $|\pi_p\rangle$ replaced by
its similar expansion, and this equation plus those governing the pion
(from the momentum operators plus the two Poincaré labeling operators), including interactions, are solved giving the c's, as well as the
corresponding coefficients for the pion. The initial conditions might be
for a single proton at rest; then the c's are trivial (but irrelevant since
if the system consisted of but a single proton there would be nothing
to describe). However the initial state might have, say, a proton and
pion moving toward each other. Solutions of these equations, for given
initial conditions, then completely describe the system; terms in the
expansion vary in space and time, as indicated. This is exact, except of
course it is impossible to solve the equations exactly. But that is not
relevant for this procedure describing how to set up the theory.

IV.2.d.ii   *The expansion for an object that can decay*

Physical object Δ_p — Δ satisfies Dirac's equation without strong interactions (thus here with no interaction) — can decay because of these
interactions into a proton (or neutron) plus a π_p. What experimentally
is the state like? The physical Δ_p has an expansion like that for the proton, except that as it decays there are extra terms (with coefficients w),
thus, suppressing sums and integrals, the expression is (schematically)

$$|\Delta_p, x, t\rangle = d_p(x,t)|\Delta_p\rangle + w(x,t)|N_p\rangle|\pi_p\rangle$$

$$= d(x,t)|\Delta) + \sum d(x,t,n)_{\Delta\pi}|\Delta)|\pi)^n + \dots$$
$$+ w(x,t)\{c_N(x,t)|N) + \sum_n c_{N,n\pi}(x,t)|N)|\pi_p)^n + \dots\}|\pi_p),$$

(IV.2.d.ii-1)

putting in the previous expression (eq. IV.2.d.i-1) for $|P_p)$ and summing over all states to which the Δ is connected by interactions, including any number of pions (allowed by conservation laws), and so on.

In the summations there is one part that involves integrals over momenta. States satisfying equations without interactions are eigenstates of momentum operators so

$$|\Delta) \sim C(p)exp(ipx) \Rightarrow \int dp exp(ipx)a^*|0),$$ (IV.2.d.ii-2)

using the creation operator formalism, but with interactions the states are sums of such terms,

$$|\Delta_p) = d(x,t)exp(ipx)|\Delta)$$
$$+ \int dqF(q)d(x,t)_{\Delta\pi}exp(iqx)exp[i(p-q)x]|\Delta)|\pi) + \dots .$$

(IV.2.d.ii-3)

Internal parts are suppressed but indicated by summations, the terms in the exponentials are scalar products in four space, so p,q,x are vectors, and there are terms like this (with more and more integrals) for each number of π's. Therefore we write

$$|\Delta,x) \Rightarrow A^*(p)|0) = |\Delta_p) = d|\Delta) + \sum d_n|\Delta)|\pi)^n + w|N_p)|\pi_p)$$
$$= c_0 exp(ipx)a^*_{\Delta}|0)$$
$$+ \sum \int dqB(p,q)exp(iqx)exp[i(p-q)x]a^*_{\Delta}(q)a^*_{\pi}(p-q)|0)$$
$$+ \dots + w|N_p)|\pi_p).$$ (IV.2.d.ii-4)

There are similar expansions for the $N\pi$ terms. These include sums over (suppressed) indices for spin and internal quantum numbers, and there is a similar term for each number of π's, as well as for each number of Δ, Δ' pairs, plus these with π's. Symbols are often schematic, so vector indices may be suppressed for example, but in actually applying the formalism all symbols and expressions must be written with a level of care that is not needed in this purely explanatory discussion.

If the object decays, or otherwise interacts with another physical object (as in scattering) the coefficients in its expansion are functions of space and time, as explicitly indicated. This dependence is often suppressed, but at times it is necessary to include it.

IV.2.d.iii *Including phases*

This expression is for a state of a single particle (really a state built upon, as given by the first term, a single group representation basis vector). It could better be written

$$|\Delta_p(p); x, x_o) = c_0 exp[ip(x - x_o)]a_\Delta^*|0)$$

$$+ \sum \int dqB(p,q)exp[iq(x-x_o)]exp[i(p-q)(x-x_o)]a_\Delta^* a_\pi^*|0)$$

$$+ \ldots, \quad \text{(IV.2.d.iii-1)}$$

so that all terms have the same phase, which is thus unmeasurable. The x_o is the same for all since it refers to the position of the physical Δ (the center of the wavepacket).

A two particle state (properly symmetrized) is then (schematically)

$$|x_i, x_j) \Rightarrow \delta(x - x_i)\delta(x - x_j)$$

$$\Rightarrow \int dp_i f(p_i)dp_j g(p_j)exp[ip_i(x-x_i)]exp[ip_j(x-x_j)]A_i^* A_j^*|0)$$

$$= exp[i(p_i + p_j)x)] \int dp_i f(p_i)dp_j g(p_j)$$

$$\times exp[-ip_i x_i]exp[-ip_j x_j]A_i^* A_j^*|0), \quad \text{(IV.2.d.iii-2)}$$

where the states are the physical ones. This has an overall space-dependent phase. However since the origin is arbitrary, this is meaningless, and can have no physical effect.

Generalization to n particles is immediate.

IV.2.d.iv *This does not apply to gravitation*

There is one interaction for which this expansion is not plausible, and that is gravitation. This expansion gives a physical state as a sum over products of free states (or in some cases ones in which a particular interaction is removed). However for gravitation that means a product of, say, an electron not subject to gravity (that is obeying Dirac's equation without the gravitational interaction) with a "graviton", but these do not exist since gravitation is nonlinear [Mirman (1995c), sec. 11.2.2, p. 187]. Thus in studying physical objects we have to use the solutions of the equations with gravity included.

IV.2.e Multiparticle states

How can we express states of a multiparticle system [Mirman (2001), sec. III.3.d, p. 127] such as one that appears in a sum for a statefunction of a decaying particle? In what way is there a difference between single particle states, and ones describing several; how can we tell from a statefunction? Discussion of this explains both physics and notation, so is given here, explaining this physically; the functional form was considered before [Mirman (2001), sec. II.5.b.iii, p. 89].

Bosons provide the most extreme case: particles with the same state — what does this mean? If two have the same energy and momentum (for wavepackets the same distribution of these so of the same form, with centers at the same point), how do we distinguish in this extreme case between a state of two particles, each with (four-)momentum p so with, we might think, statefunction $exp(i2px)$, and that of one with momentum $2p$, having (apparently) the same statefunction?

The statefunction for the pair, with

$$\sum p_\mu^2 = p_4^2 - \sum p_i^2 = m^2 \qquad \text{(IV.2.e-1)}$$

for each particle, can be written (schematically)

$$|\text{pair},p) = exp(2ip_\mu x_\mu) = exp[2i(p_4 x_4 - p_1 x_1 - p_2 x_2 - p_3 x_3)], \qquad \text{(IV.2.e-2)}$$

while for a single particle with energy $2p_4$ and three-momentum $2p_i$,

$$|\text{one},1) = exp[i(2p_4 x_4 - 2p_1 x_1 - 2p_2 x_2 - 2p_3 x_3)], \qquad \text{(IV.2.e-3)}$$

so

$$4p_4^2 - 4\sum p_i^2 = 4m^2. \qquad \text{(IV.2.e-4)}$$

Thus if we know that there exists an object with mass m, but none with mass $2m$, the first statefunction must be for a pair.

Two bosons with momentum (and charge) identical, but spin directions (slightly) different, have a state given by $a^* b^*$, where a and b are different creation operators, for different spin directions. In the limit, in which spin directions become the same, states become indistinguishable, but the formalism does not change. Thus we still use two different labels for the creation operators. Instead of writing an n-particle state as $a^{*n}|0)$, it is more explicitly written $a_1^* \ldots a_n^*|0)$, which is important in bookkeeping. But it should be clear what is meant even with an abbreviated form.

In general particles do not have exactly the same momentum, and this distinguishes states of a single object from states of multiple ones. To illustrate take states of two bosons given by a plane wave and a wavepacket. Either particle can be in either of the states so the statefunction of the system must be symmetric under their interchange. However these states are different. The statefunction of the system is thus (schematically)

$$|system) = exp(ip_\mu x_\mu) \int dq f(q) exp[iq_\mu (x_\mu - x_\mu^o)]; \qquad \text{(IV.2.e-5)}$$

x_μ^o is the center of the wavepacket. Expressing it in terms of creation operators we write

$$|system) = A^*(p) \int dq A^*(q) f(q) |0).$$ (IV.2.e-6)

Since for bosons the A's commute, this is properly symmetrized.

A general state thus has terms of the form (schematically)

$$|system, n) = \int dq_i f_i(q_i) exp[iq_{i\mu}(x_\mu - x_{i\mu}^o)]$$

$$\times \int d_j q f_j(q_j) exp[iq_{j\mu}(x_\mu - x_{j\mu}^o)] \ldots,$$ (IV.2.e-7)

with one such term for each particle (summed or integrated over any other variables). These are all functions of the same variable, x, however they are all relatively displaced, requiring the x_i^o term, different for each object. The wavepacket is, in general, different for different objects. Also each is a physical object, so that for each the statefunction given here is replaced by an expansion of a form similar to that of Δ_p (sec. IV.2.d.ii, p. 182).

Exponentiations of momentum operators translate objects, so functions describing them still depend on x (we pick some point in space, x, and measure amplitudes of statefunctions there), but centers of wavepackets are displaced by translations. There is a nonzero probability of finding an object anywhere; it varies with x_i, and time. Statefunctions are found at each point by having the (exponentiated) momentum operators act on those at the origin; momentum operators act on statefunctions at x_i^o. These are nonlinear so the set of objects given by a statefunction changes in space and time.

We can simplify the notation by dropping the x (although it is important to remember what statefunctions mean), and write them as functions of x_i^o, dropping also the superscript. Statefunction $\psi_i(x)$ then gives the probability of finding particle i at position x (which is understood) where the phase (or center of the wavepacket) is measured from a point displaced from the origin by x_i. Since the statefunction is really $\psi_i(x - x_i)$, this gives the probability for point x, and the probabilities for $\psi_i(x - x_i)$ and $\psi_{i'}(x - x_i')$ at point x differ (even though the functions of x may be the same), because the centers of the wavepackets do.

This state also differs from that of a state of one object in that the number operator has a different eigenvalue. However this is meaningful only if the value of this operator has a physically measurable effect, as say in scattering. If two objects with exactly the same four-momentum were to scatter the cross-section would be different than if one with the same four-momentum as the pair were to undergo that scattering. And although initial states have the same momentum, final states do not;

the two particles would likely come off with different momentum. This physically distinguishes states of different numbers of objects.

IV.2.f   Why these are the expressions for the states

These analyses are based on expressions for the physical states as sums of products of free ones. Why do we write them in this form?

Physical statefunctions are solutions of (many coupled) nonlinear equations thus cannot be given explicitly. How can we characterize them, at least in a formal way? The functions we have, whose forms we know, are statefunctions of free particles, those obeying linear equations. It is a fundamental assumption that these form complete sets, so that physical statefunctions can be expanded in terms of them. If this assumption were to fail it is difficult to see how we could express physical states, and if we cannot how we could study them. Thus it seems we must assume this.

What is the form of these expansions? Clearly they should be sums and integrals over the free states, over momentum, over spin, and so on. But this is insufficient, objects interact with each other and expressions for the states must be determined by their interactions. Moreover each object interacts with many different ones, and all these must be included. Thus the complete sets of states in which a physical statefunction is expanded must incorporate these states, the free states of all these other objects. But such expansions cannot simply be sums over the different objects with which an object interacts — a proton cannot be written as a sum of pions, for example.

We then must find terms all of the same type, all either fermions, or all bosons. And quantum numbers of each term should be the same, all with the same spin, the same isospin, and so on. Thus each term must be a function of the free states of the various particles with which the object interacts, and satisfy these conditions on quantum numbers. Such terms are products, and it is difficult to see how other types of terms could be acceptable, or even how there could be other types.

So we are lead to express physical statefunctions as sums of products of statefunctions of all free objects corresponding to the physical objects with which an object interacts. These seem to be the only possible terms in such expansions. We thus write a physical statefunction as a sum over all such terms, but with unknown coefficients, for if we could determine these coefficients we would exactly solve the nonlinear equations. It may be that not all of these terms appear, but this possibility is included since coefficients can be zero.

Actually determining the coefficients is really impossible, but that is a different problem from the one here, which is merely developing a formalism that can be used as a foundation for further study.

IV.2.g   Equations governing systems

Properties of a system are determined by the form of the operators acting on it, in particular the momentum operators containing the interactions. For a free object these are

$$P_\mu = i\delta^4(p_4^2 - \sum p_i^2 - m^2)\frac{d}{dx_\mu}, \qquad (\text{IV.2.g-1})$$

which does not express well the condition on the sum of the squares of the momentum eigenvalues, so better, using creation and annihilation operators (for physical objects),

$$P_\mu = \sum \int d^4p_\nu\, p_\mu A_i^*(p_\nu = p_4, p_1, p_2, p_3)A_i(p_\nu)\delta^4(p_4^2 - \sum p_i^2 - m^2)$$

$$= i\sum \int d^4p_\nu\, \delta^4(p_4^2 - \sum p_i^2 - m^2)A_i^*(p_\nu)A_i(p_\nu)\frac{d}{dx_\mu}, \qquad (\text{IV.2.g-2})$$

with indices, summed over, for sets of particles; P_4 is the Hamiltonian. With spin the operators are A_i, and so on, where the subscript gives the z-component of the spin, plus additional indices for other quantum numbers, these summed over.

Differential forms of operators are specific realizations of members of the Lie algebra of the group. Their eigenstates are restricted to those that are also eigenfunctions of the invariant operators, ones with eigenvalues obeying the relations given by these. Forms with creation and annihilation operators provide different realizations, with conditions on eigenfunctions explicit; these are nonzero only on functions for which the requirements are satisfied. On others they give zero.

A momentum operator of a free object, for the type of representations considered here, is essentially the number operator times the momentum (component) value, integrated over (the subspace determined by the representation-labeling invariants of) all momentum values, and summed (or integrated) over all other relevant indices.

IV.2.g.i   *The parameters determining the coefficients*

There are several parameters here on which coefficients depend, coupling constants and masses of free particles. Physically the latter cannot play a part; only physical states obey the equations and the masses that appear in them are physical, measured, ones. So it is the values of physical masses that we must find and use in these equations.

Except for specification of the number and types of objects and their masses (the values of p^2), and quantum numbers, this is completely general. It is also necessary to give the values of the second Poincaré invariant for these (given by subscripts), that is their spins. But that exhausts what can be given for a free particle (from the Poincaré group, so ignoring internal quantum numbers like isospin).

IV.2.g.ii   *The meaning of the sums giving the momenta*

We write physical states as sums over products of "free" states, thus with "free" operators, the a's (because these are the ones that we can postulate properties for). Momentum P is then taken to act on each state in this sum of products, thus is of the form

$$P_\mu = \int d^4 p_\nu \, p_\mu \delta^4(p_\nu^2 - m^2)\{a^*(p_\nu)a(p_\nu)$$

$$+ \int d^4 p_\rho a^*(p_\nu)a(p_\nu)a^*(p_\rho)a(p_\rho) + \ldots\}; \qquad \text{(IV.2.g.ii-1)}$$

there are an infinite number of such terms of products, containing more and more a's.

Each term acts only on the corresponding one in the sum symbolizing (giving the mathematical representation of) the state, that with the same number (and type) of a^*'s. For physical states we use for example an expansion like that for the Δ_p (eq. IV.2.d.ii-1, p. 183) which contains in its sum a term for a Δ, a term for a $\Delta\pi$, and so on, so the only term of P that contributes to the first is one with a single annihilation operator, a, to the second the term with $a_\Delta^* a_\pi^*$, and so on. Essentially then P is a sum of products of number operators (times momentum values); the first term in the expansion of the state contains no π's, so all terms in the sum with a π give zero on it, the second term in the state contains one π, and likewise for all. Thus every term in P acts only on the corresponding term in the sum for the state. So we only need the first term in P, even though it is a function of the a's, not the A's. For combinations like $N\pi$, other terms are needed. Thus P is effectively just the number operator (times p) for the system being considered, really a sum of such terms, each for a different number operator. On a product of states since momenta and spins of states are different, P acts on each individually, so translates each (but as a sum translates them simultaneously and, exponentiated, the same distance).

IV.2.g.iii   *Inclusion of interactions*

Momentum operators are nonlinear, they include interactions. Thus we have to define how nonlinear terms act on states. A system is specified by the quantum numbers, masses and interactions of the objects of which it consists. Here as an example we assume a single interaction, writing the interaction Hamiltonian as $g_{\Delta\pi}|\pi_p\rangle|\Delta_p\rangle\langle\Delta_p|$, to indicate the interaction of a Δ with a π, so expanding these states, schematically,

$$H_{int} = g_{\Delta\pi}|\pi_p\rangle|\Delta_p\rangle\langle\Delta_p| + \text{hermitian conjugate}$$

$$\Rightarrow g_{\Delta\pi} \int d^4 p_\mu d^4 q_\nu d^4 r_\sigma C_0 \delta^4(p_\mu - q_\nu - r_\sigma)$$

$$\times \delta(p^2 - m_\Delta^2)\langle 0|\{A_\Delta(p_\mu)A_\Delta^*(q_\nu)A_\pi^*(r_\sigma)$$

$$+A_\Delta(q_v)A_\pi(r_\sigma)A_\Delta(p_\mu)^*\}|0). \quad \text{(IV.2.g.iii–1)}$$

Suppressed are spin indices and sums over them; Clebsch-Gordan coefficients needed for H to be a (three-)scalar are merely indicated by C_o (which for spin-$\frac{1}{2}$ objects might be given by Dirac γ's). The C_μ's denote the proper Clebsch-Gordan coefficients for the P_μ's; the ones for different the μ's are related by Lorentz transformations. This operator, acting on a state, annihilates a Δ and creates a $\Delta\pi$ pair, and also annihilates such a pair and creates a Δ. These are physical operators, which means that if particles also interact with others, they are included in the operators. Thus $A_\Delta(q_v)$ annihilates a physical Δ_p when acting on a state, so if there is an interaction between a Δ and an $N\pi$, the state it annihilates is a sum of the Δ plus a term for the $N\pi$ pair, plus others, and similarly for the rest of the operators.

Therefore, acting on a physical Δ with momentum p (schematically),

$$P_\mu(\Delta)|\Delta_p,p) = \int d^4p_v p_\mu A^*(p_v)A(p_v)\delta(p2-m^2)|\Delta_p,p)$$

$$+g_{\Delta\pi}\int d^4p_v C_\mu A^*(p_v)A(p_v)\delta(p^2-m^2)|\pi_p)|\Delta_p,p),$$

$$\text{(IV.2.g.iii–2)}$$

and substituting the expressions for the physical states (eq. IV.2.d.ii–4, p. 183),

$$P_\mu(\Delta)|\Delta_p,p) = \int d^4p_v p_\mu \delta(p^2-m^2)a^*(p_v)a(p_v)d(x)a_\Delta^*(p)|0) +$$

$$\sum\int d^4q B(q,x)a^*(p_v)a(p_v)a_\Delta^*(p-q)a_\pi^*(q)|0) + \ldots + \sum d_n|N)|\pi)^n$$

$$+\ldots+g_{\Delta\pi}d^4p_v\int p_\mu dq_v dr_\sigma C_\mu \delta(p^2-m^2)\delta^4(p_\mu-p_v-q_v-r_\sigma)$$

$$\times A^*(p_v)A(p_v)A_\Delta^*(q_v)A_\pi^*(r_\sigma)|0) + \ldots. \quad \text{(IV.2.g.iii–3)}$$

This is an iterative process; we do not know the A's, so have to use the expressions for the expansion of the states for them in the interaction terms. Physical operators are sums over the (nonphysical) free ones with coefficients that we do not know, but have to find by solving the equations, and these depend on the unknown coefficients.

IV.2.g.iv All objects must be included in interactions

Every object interacts with many others. Thus each baryon interacts with (apparently) each type of meson, of which there are many, perhaps an infinite number, plus (through these) other baryons, and directly or indirectly photons, and so on. In the Hamiltonian for the Δ, which interacts with pions, omegas, nucleons (denoted by N) there is thus a term for the πN interaction, another for ωN, one for $K\Lambda$, and so on,

$$H_{int} = + \ldots g_{\Delta N \pi}(\Delta|\pi_p)|N_p) + hc + \ldots = \ldots$$

$$+ g_{\Delta N \pi} \int d^4 p_\mu d^4 q_\mu d^4 r_\mu \delta^4(p - q - r) C_4 \{A_\Delta(p_\mu)A_N^*(q_\mu)A_\pi^*(r_\mu)$$

$$+ A_N(q_\mu)A_\pi(r_\mu)A_\Delta(p_\mu)^*\}|0) + \ldots . \quad \text{(IV.2.g.iv-1)}$$

This is one term of the perhaps infinite number, and there should be indices, and perhaps matrices, for spin and internal quantum numbers, these summed over. Also this is one component of the four-momentum, and that should be indicated by a subscript on a statefunction or on a matrix (here one is placed on the coefficient), and similarly for the other momentum components (these are related, their subscripts are related, by Lorentz transformations). Each state appearing here, the π, the N and so on, is governed by a similar set of four-momentum (and representation-labeling) operators. Thus a state of the system is a product of states of each of these particles, and the momentum operators are sums of those for each particle type.

IV.2.g.v   *How an infinite number of pions occur*

Why in this expansion are there an infinite number of pions, that is terms $|\Delta)|\pi)^r$, for all $r \geq 0$? The interaction Hamiltonian is (schematically)

$$H_{int} = \sum_r A^* A(B + B^*)^r; \quad \text{(IV.2.g.v-1)}$$

A is the annihilation operator for $|\Delta)$, A^* the creation operator, with B's the corresponding pion operators. This, acting on $|\Delta)|\pi)^r$, gives terms $|\Delta)|\pi)^{r+1}$ and $|\Delta)|\pi)^{r-1}$, so the expansion includes all $r \geq 0$.

We can consider general Hamiltonians in which there are many r's, or in which there is no limit on them. Also the displacement is given by $exp(ip_4x_4)$, for time-translation operator $exp(iHt)$, and expanding gives all powers of the momentum (including the Hamiltonian) so all powers r.

IV.2.g.vi   *Equations must transform properly, so restrict interactions*

Interactions are strongly restricted by the requirement that all terms in an equation transform the same way under the geometrical transformation group, as is well-known. In particular, the interactions with massless objects, especially the electromagnetic field, are completely specified, giving minimal coupling [Mirman (1995c), sec. 4.2, p. 57].

It might seem that this does not always hold [Ohanian and Ruffini (1994), p. 380], for example because of the anomalous magnetic moments of nucleons. However this again emphasizes that such objects do not obey not the free Dirac equation with only the electromagnetic

interaction, but one with other interactions, most important here the
strong ones [Mirman (2001), sec. V.2.d, p. 205]. Thus magnetic moments calculated from Dirac's equation with only minimal coupling of
the electromagnetic interaction have to be wrong. The anomalous part
merely reflects the changes required by the other terms, the other interactions, in the correct equation. These cannot imply the absence of
minimal coupling.

IV.2.h   The Dirac equation and its consequences

Dirac's equation is the best known of those governing particles, so it is
useful to summarize how these considerations apply to it (even though
it is only for a special case, spin-$\frac{1}{2}$). For Δ_p it is (with other interactions,
and integrals and sums, suppressed)

$$iy_\mu \frac{d|\Delta_p)}{dx_\mu} + \{g_{\Delta\pi}|\pi_p)|\Delta_p)(\Delta_p| + hc\}$$

$$+ \{g_{N\pi}|\pi_p)|N_p)(\Delta_p| + hc\} - M_P|\Delta_p) = 0; \qquad \text{(IV.2.h-1)}$$

this includes the possibility of an interaction between the Δ and π which
gives elastic scattering, as well as between N and π resulting in decay
of the Δ and change of particle type. For Δ,

$$iy_\mu \frac{d|\Delta)}{dx_\mu} - m|\Delta) = 0; \qquad \text{(IV.2.h-2)}$$

m is a parameter whose value is not determinable (and irrelevant).

We need an expression for $|\Delta_p)$. With no interactions it is just $|\Delta)$,
and in general a sum including this as the first term. The expression
for this state must be given in terms of the objects of the formalism,
there being no others, the creation and annihilation operators for the
free particles. Thus, indicating just two of the terms,

$$|\Delta_p) = c_0|\Delta) + \sum c_n|\Delta)|\pi)^n + \sum d_n|N)|\pi)^n; \qquad \text{(IV.2.h-3)}$$

the sum is over all number of π's, all n's, from 1 to infinity. Summations
also represent sums over internal labels and integrals over momenta.
These, and all other irrelevancies, like spin and coordinates, are suppressed (although they must be considered explicitly for specific cases).
That they can be suppressed is a consequence of the generality of this
discussion.

The terms in the first sum are all "virtual", they are states obeying the
free particle equations, so with no direct physical meaning, but some in
the second can be replaced with "real" particles, since the Δ can decay.
We have equations for all types of these objects, and expansions of this
type for all. These expansions are now substituted in the equations and

they are solved (in principle) for the coefficients, giving the physical states. We also require normalization, say,

$$(\Delta_p | \Delta_p) = 1, \qquad \text{(IV.2.h-4)}$$

and likewise for the others. In addition for decay or scattering we start with an initial state, initial values of the c's and d's that are consistent with the requirements, and find how they develop in time.

IV.2.i   These discussions are purely formal

Of course finding expressions for states obeying these nonlinear equations is not likely to be possible, but that is irrelevant for our purpose, the study of the nature of expressions and properties of these states (the formalism), and requirements that symmetry places on them. Although we represent objects by infinite expansions, the arguments are exact. We do not calculate so need not truncate, using series only to express and find properties of statefunctions. While there are questions about the convergence of infinite series, at this point they do not arise since these are purely formal. Of course the theory of nonlinear realizations of groups is not well developed and there are undoubtedly aspects, including perhaps some that arise here, that need a more thorough examination to determine if they are rigorously correct. However since we are primarily interested in introducing the methods, expressions, and nomenclature, and seeing how the conformal group affects the theory, we do not consider these more subtle parts.

But perhaps study of the formalism will stimulate work on them.

IV.2.j   Is quantum mechanics linear?

These expressions allow us to consider whether quantum mechanics is linear. The answer depends on the formalism [Mirman (2001), sec. II.4.b, p. 81]. If we write Dirac's equation with interactions, schematically,

$$i\gamma_\mu d_\mu \psi - m\psi + g\psi\psi^* \Delta = 0, \qquad \text{(IV.2.j-1)}$$

and similar equations for Δ, then it clearly is not linear. The nonlinearities, interactions, are essential. A sum of solutions is not a solution. The statefunction for two charged particles, say, is not the sum of statefunctions of each — electromagnetic interactions prevent that.

However if we write a statevector, say for a decaying object, schematically although the argument is general,

$$|s; x, t) = c(x, t)_\Delta |\Delta) + c(x, t)_{N\pi} |N, \pi) + \ldots, \qquad \text{(IV.2.j-2)}$$

then the momentum operators, most familiarly the Hamiltonian, give

$$-i\frac{d|s;x,t)}{dt} = H|s;x,t) = -i\frac{dc(x,t)_\Delta}{dt}|\Delta) - i\frac{dc(x,t)_{N\pi}}{dt}|\Delta,\pi) + \ldots,$$

(IV.2.j-3)

and similarly for the other momentum operators. In this form it is linear. Here the nonlinearities have been transferred from the equations to the momentum operators. In this case a sum of solutions is a solution. But this is only because the terms making up the sum do not describe objects interacting with each other. They would be for, say, two Δ particles, each of which interacts with π's (which do not interact), but not with each other.

Although it is a fundamental postulate, for which arguments can be given [Caticha (1998a,b)], that quantum mechanics is linear, this is true only because of the way we treat it. It is however a property of some of the groups, like the rotation group, but not necessarily the Poincaré group [Mirman (1995c), sec. 1.2.1, p. 7], that sums of basis states are basis states, which requires linearity, but with a different meaning for linear — such sums are not for statefunctions of different particles.

For these groups the statefunction of a set of noninteracting objects is a sum of statefunctions of each. A basis vector of the rotation group with spin along the z' axis is a sum of basis vectors along the z and y axes. However while each of these can serve as statefunctions they do not if there is a single object having spin along z'. Rather the sum is a different way of writing that statefunction, useful if the z and y axes are chosen for other purposes.

This emphasizes again that obvious statements may not actually be so if terms are required to be explicit. It is quite easy to use language that is ambiguous [Mirman (2001), chap. IV, p. 142].

IV.3   Requirements from transformation groups

With this foundation for statefunctions (of quantum field theories) and operators (of transformation groups) on them we can now consider how these groups, the rotation, Lorentz, Poincaré groups, the conformal group, and perhaps larger ones, place conditions on the formalism, so on physics, and what information they provide.

Requirements due to the Poincaré group and its subgroups are well-known [Kim and Noz (1986); Mirman (1995c); Sudarshan and Mukunda (1983), p. 439], so we only briefly review them.

IV.3.a   The effect of the Poincaré group

First we consider conditions placed by the Poincaré subgroup of the conformal group. We use (for noninteracting objects) the realization

$$P_\mu = i\frac{d}{dx_\mu},\tag{IV.3.a-1}$$

and

$$M_{\mu\nu} = i(x_\mu d_\nu - x_\nu d_\mu) + i\Sigma_{\mu\nu},\tag{IV.3.a-2}$$

so states (momentum eigenstates) are of the form

$$|x,t,j) = \int d\omega dk c(k,\omega,j) exp(ikx + i\omega t),\tag{IV.3.a-3}$$

where kx is the scalar product, and j is the internal (here spin) label.

Note that Σ commutes with the P's so commutation relations are satisfied whether Σ is present or not; commutation relations with the P's do not place restrictions on it. Its realization is given by the nature of the states on which it acts.

IV.3.a.i   *Realization of the internal angular momentum generator*

How is Lie algebra internal operator $\Sigma_{\mu\nu}$ realized acting on a free particle, thus on a $\Delta(p_\nu)_m$, with m the spin label? For $\Sigma_{ij}, i,j = 1\ldots3$,

$$\Sigma_{ij}a^*_{\Delta m} \Rightarrow C_{mn}(ij)a^*_{\Delta n},\tag{IV.3.a.i-1}$$

where C is the relevant matrix element. Hence

$$\Sigma_{ij} = \Sigma C(ij,mn)a^*_{\Delta n}a_{\Delta m},\tag{IV.3.a.i-2}$$

summed over all values of the indices (for some, matrix elements may be zero). The effect of spin operators is well-known: $a^*_{\Delta u}$ creates a particle with spin up along z; Σ_{xy} changes it to a creation operator with spin unchanged, so is a rotation around z, thus giving only a phase change. Generators

$$\Sigma_+ = \frac{1}{2}(\Sigma_{xz} + \Sigma_{yz}) = \sum_m C(+,m,m+1)a^*_{\Delta m+1}a_{\Delta m},\tag{IV.3.a.i-3}$$

and

$$\Sigma_- = \frac{1}{2}(\Sigma_{xz} - \Sigma_{yz}) = \sum_m C(-,m,m-1)a^*_{\Delta m-1}a_{\Delta m},\tag{IV.3.a.i-4}$$

change the direction of spin, that is take a creation operator for one spin direction to another with the spin component changed by ±1.

IV.3.a.ii *Boost generators*

More interesting is

$$\Sigma_{4z} = \Sigma C(4z, mn) a^*_{\Delta n} a_{\Delta m},\qquad\qquad (\text{IV.3.a.ii–1})$$

which goes with a boost along z, so leaving the spin component unchanged. To study operators

$$\Sigma_{4\pm} = \frac{1}{2}(\Sigma_{4x} \pm \Sigma_{4y}) \qquad\qquad (\text{IV.3.a.ii–2})$$

we use the matrix elements of the Lorentz group generators [Gel'fand, Minlos and Shapiro (1963), p. 188; Naimark (1964), p. 116]. Representations are labeled by two integers which, for finite-dimensional representations, give the minimum and maximum angular momenta (sec. A.2, p. 224). For spin-$\frac{1}{2}$ the maximum and minimum are equal.

However in general the total spin, as well as its component, are changed (sec. A.4, p. 233). We might think that an object with intrinsic spin-$\frac{3}{2}$ at rest, when moving, has spin that is a combination of spin-$\frac{3}{2}$, spin-$\frac{5}{2}$ and spin-$\frac{7}{2}$. But we cannot consider (except in special cases) intrinsic spin of a moving object. At rest the total angular momentum of an object equals its spin. It is incorrect to distribute angular momentum of a moving object into spin and orbital angular momentum. It is the total that has meaning.

Subscript m in $a_{\Delta m}$ here runs over all su(2) representations and states in the Lorentz group representation. State

$$|\Delta, p_v, m, x\rangle = exp(ip_v x_v) a^*_{\Delta m}|0\rangle, \qquad\qquad (\text{IV.3.a.ii–3})$$

has momentum p_v and spin m, this being the total spin (but of course not the total angular momentum) and its component. A boost operator changes this to a state with a different momentum which is also a sum of states of different total angular momentum and components.

So, with t, s labeling the total angular momentum,

$$\Sigma_{4\pm} \sim \Sigma\pm \Rightarrow \Sigma_{4\pm} = \sum_{t,s}\sum_{m} C(t, s, \pm, m, m\pm 1)' a^*_{\Delta t, m\pm 1} a_{\Delta s, m}, \quad (\text{IV.3.a.ii–4})$$

$$\Sigma_{4} \sim \Sigma_{4z} \Rightarrow \Sigma_{4} = \sum_{t,s}\sum_{m} C(t, s, 4, m, m)' a^*_{\Delta t, m} a_{\Delta s, m}, \qquad (\text{IV.3.a.ii–5})$$

and these act on spin the same way that rotation operators do, but change the phase of the state differently. Hence Σ_{4x} going with a boost along x does change the spin direction and except in special cases the total spin (appendix A, p. 223).

Thus we have conditions on both statefunctions, the scalar coefficients for example, and on the realizations of the operators, these following in part from the assumed realization of the P's.

IV.3.a.iii   *Spin and orbital angular momentum are thus related*

That a moving object does not, in general, have a unique spin has an important implication. There is a fundamental relationship between orbital angular momentum and spin for they are only parts of what is the meaningful quantity. When a rotation changes the direction of one, it must act on the other also. They cannot be separated. As we would expect then (and as we know), it is total angular momentum that is conserved, rather than each part alone.

Why does a rotation of coordinates induce one of spin? The Poincaré group requires it [Mirman (1995b), sec. 6.3, p. 114]. Also the Hamiltonian links spin and orbital angular momentum. If these were not linked then conservation of angular momentum would involve only the orbital part, so we could not say that spin is a form of angular momentum but would have to regard it as a purely internal label, like isospin.

While particles like the electron or nucleon are sometimes pictured as spinning tops, this clearly can only be a visualization. Spin is a form of angular momentum because rotational invariance requires that total angular momentum be conserved (the reason that angular momentum has physical meaning). But neither spin nor orbital angular momentum are separately conserved in general, meaning that spin is angular momentum.

IV.3.a.iv   *Implications of the interrelationship of spin and orbital angular momentum*

The relationship between spin and orbital angular momentum has other consequences. Consider a hydrogen atom, with the proton the observer [Mirman (1995b), sec. 5.1.3, p. 88], the electron the object. Observations by the proton, that is the effects of the electron on it, depend on the orientations of the spin and orbital angular momentum of the electron with respect to the spin of the proton. We can rotate the observer, the proton (carefully, without perturbing the system), changing its spin from up to down. The observations made by it — its interactions with the electron — are now different. We can regard the rotation to also rotate the spin of the electron, and the direction of its orbital angular momentum, thus of the space coordinates. Or we can consider an electron moving along the z axis, defined by the proton's spin, with the rotation reversing the direction of the electron's momentum. Thus a rotation of space (say changing the direction of the orbital angular momentum) induces that of spin, a rotation of spin induces one of space (specifically, a rotation of the spin of the observer changes the coordinate system that it sees, so that it regards the coordinates, thus orbital angular momentum, to have been rotated). Rotations of coordinates refer to the relative change of the external world with respect to the intrinsic coordinate system of the observer.

Another example is given by two observers, say in a helium atom, with the two protons the observers. If they have opposite spin, then their observations — their interactions with an electron — differ because both its spin and its orbital angular momentum directions are different for the two observers. A rotation from one observer, from one proton to the other, rotates, in that sense, both the spin of the electron and its orbital angular momentum, and direction of motion.

Why can objects, say protons, with different orbital angular momentum be created? Equations governing them, the Dirac equation for example, contain derivatives with respect to space. Interaction terms contain y_μ thus act on spin, and can contain operators like τ_ν acting on isospin. But they do not contain operators acting on internal space-like variables. And since these (initial) statefunctions do not depend on internal space-like variables, the final ones cannot — no statefunctions can. Therefore such variables are not observable, so do not exist. Also interactions are products of statefunctions of different objects, these space dependent. Thus in scattering, say, final statefunctions can have different space dependence than initial ones. Orbital angular momentum can therefore be changed (and there is no such thing as internal orbital angular momentum).

While a basis vector may appear in the representation space of a group, it may not exist physically. A statefunction of an object is obtained from an initial one by the action of momentum operators, including the time-translation generator, the Hamiltonian. If this does not change a basis vector, then that basis vector is the same for all objects with the same initial statefunction, and presumably thus for all. Variables can be observed only if the Hamiltonian (and so because of the Lorentz group, all four-momentum operators) change them. That there is a basis vector in the representation space of the homogeneous subgroup does not mean that it can be reached — this is only possible if the inhomogeneous operators contain terms including it (sec. I.6.b, p. 53).

IV.3.b   Interactions and internal symmetry

For internal symmetry, we consider an "atom" with a proton moving around a "nucleus" consisting of a Δ^-, a Σ^-, say, these being the observers, or a nucleus consisting of more than one, giving different observers with different spins, directions of spin, isospins, SU(3) quantum numbers. Each makes different observations of the proton, and there are operators taking one observer to another — correlating their observations. There are operators that take a Δ to a Σ, mixing thus spin direction, spin magnitude, isospin and SU(3) quantum numbers. Operators changing spin direction must also change that of orbital angular momentum. Thus these are products of operators on spin states with those acting on space coordinates.

IV.3.b.i *How su(6) illustrates these concepts*

There are indications, although far from conclusive, that particles belong to su(6) multiplets (or at least those of the algebra, which is denoted by small letters, with capitals indicating the group). While this may not be correct it can be used to illustrate many of the present concepts. The relevant decomposition is noncanonical, su(6) ⊃ su(2) ⊗ su(3). These ideas can also be illustrated, somewhat less completely, by the more familiar su(3), for which, being so simple, the canonical and noncanonical decompositions are the same: su(3) ⊃ su(2) ⊗ su(1), with su(2) the algebra of the transformations on spin and space, while su(1) is the algebra of the internal group. The group transformations belong to the 35-dimensional representation of su(6), and the 8-dimensional one of su(3), and for both the su(2) operators belong to its 3-dimensional representation.

The algebras of these groups can be considered as purely spectrum-generating algebras, but this is somewhat artificial since particles not only belong to su(2) representations, but its operators also transform them. Could the same be true for su(6)?

If we consider particle states to be of SU(6) representations, with the group in addition transforming them, there are in these sets generators acting only on spin and space. But others act on both spin and su(3) internal coordinates whose commutators close on pure spin operators. Since spin transformations induce ones on space, the former must include operators that close on operators for space transformations. Thus while the representation space of su(3) is a (triple) direct product of ones with spin, and with internal symmetry coordinates and another with space coordinates, that of su(6) is not.

Representation spaces of the product group are products of spaces, one containing spin and isospin functions among its basis vectors, the other space and space-like functions. That part containing spin and isospin is a direct product of the two. Thus nucleon statefunctions, considering only these two parts, are four-component objects, two giving the spin of the neutron or proton, the other two, for fixed spin, labeling particle type. The rotation group mixes the first pair (plus space coordinates), the isospin group the second. But a group containing (and larger than) these two subgroups, like SU(6), then also has operators mixing pairs, mixing any pair of the four states — there are operators that change the components of spin and of isospin, but also ones changing their magnitude.

The problem this leads to is that (apparently) there is no internal space which could be acted upon by the space part of the operators, and these do exist since SU(2) operators are products of ones acting on

space, with others acting on spin. And it is not possible for the space
and spin operators to form different algebras. Since there are eight
su(6) operators not having space parts (and one for su(3)), the space
parts would have to form a 27-dimensional algebra (or a 7-dimensional
one for su(3)), and there are none.

Consider particles belonging to the 35-dimensional representation
of su(6). Could these just belong to a product of representations of
su(2) and su(3)? While su(2) has both five and seven dimensional representations, su(3) has neither. Thus there is no way of getting such a
product. Of course we do label states by their representations of these
two groups. But this su(6) representation has more than one of these
representations. So additional labels are needed (as is clear since it
has more labels than the product). Thus there are su(6) operators that
mix states that are identical under su(2) ⊗ su(3), and also states within
a subalgebra multiplet. Those acting on spin must also act on space
coordinates, some of which do not exist.

IV.3.b.ii *What transforms statefunctions?*

This raises the question of how these transformations are carried out,
and whether they must be. Objects are not transformed by (homogeneous) group operators (sec. I.5.b, p. 44), but by Hamiltonians. Statefunctions are transformed (say rotated) because there are in Hamiltonians nonlinear terms that do so. An interaction term would be (schematically) say a product of statefunctions $\psi\chi$, which are expanded in complete sets of states, and these would include products of all spin states,
with all orbital angular momentum states, say spherical harmonics, reduced so the Hamiltonian is a rotational scalar (and the fourth component of a four-vector). Each such term acting on a statefunction gives
another statefunction, say one with orbital angular momentum 1, and
up, and spin-$\frac{1}{2}$ down, to another with the component of the space part
zero, and spin up. Thus the interaction term acting on a statefunction
gives a sum of all possible statefunctions for which the total angular
momentum is conserved.

For su(6) there would be the products of each of these, with all internal symmetry states. These include terms changing the "internal orbital
angular momentum", and that is the problem. It may be that there really is such a thing, but for nonzero values, the corresponding masses
are so much greater that it has never been discovered, or perhaps it
has been but there is no way of discerning that. However this would
run into very serious problems [Mirman (1995b), chap. 7, p. 122]. (Of
course it could be that su(6) is physically incorrect.)

IV.3.b.iii *Not all transformations need be physically possible*

A here more realistic possibility is these unwanted transformations
need not be possible. Why must the Hamiltonian include states with
spin up and also with spin down? The reason is that space is invariant under these transformations, thus so must the Hamiltonian be, and
this requires both space components, and in general that the terms
in Hamiltonians be expandable in a complete set of states, like those
with all spins and all orbital angular momenta. However we can consider that for "internal space" Hamiltonians are not expandable in a
complete set of states, including ones with nonzero "internal orbital
angular momenta".

The fundamental points are that objects are not transformed by
Lorentz group operations, but by Hamiltonians, and that because a set
of objects belongs to a group representation does not say that all of
them need be observable (so not need exist) if Hamiltonians do not
contain terms reaching them from observable objects. It is not necessarily true that all transformations of a group are physically possible
(sec. I.6.b.ii, p. 54), and they are not if no Hamiltonian (can) contain
them.

IV.3.c Restrictions due to dilatations

Besides the Poincaré subgroup the conformal group contains other elements and these also place requirements on physical theories. To
investigate additional requirements we start with the similitude group,
adding dilatations to the Poincaré group. Does dilatation invariance
(sec. I.2.a, p. 9) restrict physics?

To set up a system of units we pick a mass m (arbitrarily), or length.
The statefunction of an object with this mass, in its rest frame, is proportional to $exp(ip_4t) = exp(imt)$. We ignore 2π, as not relevant
here; in actually setting up a system of units decisions have to be made
where to put it, but we are not explicitly doing so, just illustrating the
meaning of terms. The unit of time, since (sec. I.2.a.v, p. 12)

$$m_1 t_1 = 1, \text{ is } t_1 = \frac{1}{m_1}. \tag{IV.3.c-1}$$

We next take the distance that light travels in time t_1 as the unit length
($c = 1$), the Compton wavelength of m_1,

$$\lambda_1 = \frac{1}{m_1}. \tag{IV.3.c-2}$$

We now have a system of units.

IV.3.c.i   *Transformation of states under dilatations*

Transformations of coordinates and their dual, momenta, under dilatations are determined by the definition of the transformation, this being defined by its action on space. Statefunctions are basis vectors
of the similitude group. How do they transform? Groups like rotation,
Lorentz and Poincaré ones determine the transformations of their basis
vectors. Matrices acting on them have to obey the same multiplication
rules as the group elements. But here the dilatation acts on momentum,
and inversely on coordinates so px is invariant (sec. I.2.a.v, p. 12). A
statefunction transforms so that

$$\psi = |px) \Rightarrow \rho|px) \Rightarrow \rho\psi, \qquad\qquad (\text{IV.3.c.i--1})$$

and we require ρ. The only condition comes from the commutation relations between the dilatation and the momentum operators, and since
these act on only px, they are satisfied for all ρ. Other groups determine basis vectors up to normalization, and this transformation is a
change of normalization, so the freedom is not unexpected. We need
then other conditions, and require especially that they be consistent.

Statefunctions give probability densities which are changed if the
scale of coordinates is,

$$x \Rightarrow \frac{x}{\lambda}, \qquad\qquad (\text{IV.3.c.i--2})$$

thus are multiplied by a factor, and this is the same for all objects. It
does not depend on mass, or other properties. A region within space
can be marked off, and with physics invariant under dilatations, the
probability of finding an object within the markers is unchanged. This
gives one condition.

The probability of finding an object in a volume is

$$Pb(x) = \psi^*(x)\psi(x)dx^3 \Rightarrow Pb(x) = \rho^2\lambda^{-3}\psi^*(x)\psi(x)dx^3.$$
$$(\text{IV.3.c.i--3})$$

For this to be invariant under dilatations,

$$\rho^2 = \lambda^3; \quad \rho = \lambda^{\frac{3}{2}}, \qquad\qquad (\text{IV.3.c.i--4})$$

so

$$\psi' = \lambda^{\frac{3}{2}}\psi. \qquad\qquad (\text{IV.3.c.i--5})$$

This relates the transformation of basis vectors to that of coordinates,
that is their units, and agrees with the standard expression for fermions
[Schweber (1962), p. 219]. This transformation property of ψ depends,
of course, on the dimension of space. The effect of dilatations is different in different spaces. But we do live in a space of dimension 3+1, so
the space part has dimension 3.

Bosons transform differently [Schweber (1962), p. 185], a result of the definitions of the statefunctions. For these

$$\phi' = \sigma\phi \quad \text{which we see is} \quad \phi' = \lambda\phi. \tag{IV.3.c.i-6}$$

From the form of the Hamiltonians the energy for bosons is

$$E_b \sim \mu^2\phi^2 dx^3, \quad \text{so} \quad \lambda E_b \sim \lambda^2\mu^2\sigma^2\phi^2\lambda^{-3}dx^3, \tag{IV.3.c.i-7}$$

with mass μ, while for fermions, with mass m,

$$E_f \sim m\psi^2 dx^3, \quad \text{so} \quad \lambda E_f \sim \lambda m\rho^2\psi^2\lambda^{-3}dx^3. \tag{IV.3.c.i-8}$$

Hence

$$\sigma = \lambda, \quad \rho = \lambda^{\frac{3}{2}}, \tag{IV.3.c.i-9}$$

The probability densities are then compared using

$$pb_f = \psi^*(x)\psi(x) \sim \frac{E_f}{m}, \tag{IV.3.c.i-10}$$

$$pb_b = \phi^2 \sim \frac{E_b}{\mu^2}, \tag{IV.3.c.i-11}$$

and the probabilities are

$$Pb_f \sim pb_f dx^3 \sim \psi^*(x)\psi(x)dx^3 \Rightarrow \lambda^3\psi^*(x)\psi(x)\lambda^{-3}dx^3, \tag{IV.3.c.i-12}$$
$$Pb_b \sim pb_b dx^3 \sim \mu\phi^2 dx^3 \Rightarrow \lambda\mu(\lambda\phi)^2\lambda^{-3}dx^3, \tag{IV.3.c.i-13}$$

and are invariant. Thus probability density is related to energy density differently for the two particle types. But this is merely a result of the way statefunctions are defined. Bosons can be made to transform the same as fermions (or conversely) by defining their statefunctions as

$$\eta = \mu^{\frac{1}{2}}\phi. \tag{IV.3.c.i-14}$$

IV.3.c.ii *Dilatations and Dirac's equations with interacting objects*

To study further the effect of dilatation invariance, and illustrate that only mass ratios are relevant, we consider Dirac's equation with interactions, and require at least two objects. Although fermions do not seem to be directly coupled to each other, can they be?

Dirac's equations are, where ψ^*, χ^* are the hermitian conjugates,

$$i\gamma_\mu \frac{d\psi(x)}{dx_\mu} - M\psi + g\chi\chi^*\psi = 0, \tag{IV.3.c.ii-1}$$

$$i\gamma_\mu \frac{d\chi(x)}{dx_\mu} - m\chi + h\psi\psi^*\chi = 0. \tag{IV.3.c.ii-2}$$

Dilatation λ changes the equations to, with χ', ψ', the transformed solutions,

$$\lambda i\gamma_\mu \frac{d\psi(\lambda x)'}{dx_\mu} - \lambda M\psi' + g\chi'\chi^{*'}\psi' = 0, \qquad \text{(IV.3.c.ii–3)}$$

$$\lambda i\gamma_\mu \frac{d\chi(\lambda x)'}{dx_\mu} - \lambda m\chi' + h\psi'\psi^{*'}\chi' = 0. \qquad \text{(IV.3.c.ii–4)}$$

Both masses are multiplied by the same factor — they are ratios with respect to an arbitrary unit mass. The coupling constants are invariant (once this arbitrary unit mass is fixed) — their values are experimentally determinable and if they were not constants there could not be invariance (gravitation (sec. IV.3.e, p. 209) does not obey equations like these). We then get

$$\lambda i\gamma_\mu \frac{d\psi(\lambda x)}{dx_\mu} - \lambda M\psi + \lambda^{\frac{6}{2}} g\chi\chi^*\psi = 0. \qquad \text{(IV.3.c.ii–5)}$$

For the interaction term to be invariant,

$$\lambda = 1, \qquad \text{(IV.3.c.ii–6)}$$

so that this interaction is not invariant under dilatations. We conclude that such direct fermion interactions are not possible. It is clear why the weak interaction requires an intermediate boson.

However for an interaction with a boson

$$i\gamma_\mu \frac{d\psi(\lambda x)}{dx_\mu} - M\psi + g\pi\psi = 0, \qquad \text{(IV.3.c.ii–7)}$$

$$\lambda i\gamma_\mu \frac{d\psi(\lambda x)'}{dx_\mu} - \lambda M\psi' + g\pi'\psi' = 0, \qquad \text{(IV.3.c.ii–8)}$$

$$\lambda i\gamma_\mu \frac{d\psi(\lambda x)}{dx_\mu} - \lambda M\psi + \lambda g\pi\psi = 0, \qquad \text{(IV.3.c.ii–9)}$$

so this interaction is allowed by dilatation invariance.

Suppose that the boson field is defined so it transforms the same as fermion fields? Using η the equation is

$$i\gamma_\mu \frac{d\psi(x)}{dx_\mu} - M\psi + g\eta\psi = 0, \qquad \text{(IV.3.c.ii–10)}$$

$$\lambda i\gamma_\mu \frac{d\psi(\lambda x)'}{dx_\mu} - \lambda M\psi' + g'\eta'\psi' = 0, \qquad \text{(IV.3.c.ii–11)}$$

$$\lambda i\gamma_\mu \frac{d\psi(\lambda x)'}{dx_\mu} - \lambda M\psi + \lambda^{\frac{3}{2}} g'\eta'\psi = 0. \qquad \text{(IV.3.c.ii–12)}$$

Now

$$\eta = \mu^{\frac{1}{2}}\phi, \qquad\qquad\text{(IV.3.c.ii–13)}$$

so

$$\lambda^{\frac{3}{2}}g'\mu^{\frac{1}{2}} = 1, \Rightarrow g' = \lambda^{-2}g. \qquad\qquad\text{(IV.3.c.ii–14)}$$

The coupling constant would have to have units. This is a reason for the standard definition of boson fields.

This can also be seen from the Klein-Gordon equation for the boson,

$$(\frac{d^2}{dx_\mu^2} - m^2)\phi = g\psi\psi^*. \qquad\qquad\text{(IV.3.c.ii–15)}$$

Thus under dilatations

$$\psi' \sim \lambda^{\frac{3}{2}}\psi, \quad \phi' \sim \lambda\phi. \qquad\qquad\text{(IV.3.c.ii–16)}$$

From the rotation group fermion fields must enter equations for bosons in the form $\psi^*\psi$, while boson fields appear linearly. They thus transform differently so have somewhat different definitions. The transformation properties from this equation and from Dirac's equation are the same (fortunately). Also for the electromagnetic coupling, the coupling constants in these two equations are the same. Presumably this is true also for other interactions, but this is very difficult to test experimentally.

Another possibility is derivative coupling, so the interaction term is $g\gamma_\mu\frac{d\phi}{dx_\mu}\psi$. Again this requires that g have units which makes derivative coupling implausible.

Thus dilatation invariance imposes conditions on interactions, perhaps not surprisingly.

This emphasizes is that dilatation invariance is possible even though objects have mass. There are conditions that it leads to, but masslessness is not one. However it is essential that all masses transform the same and inverse to coordinates, that statefunctions transform correctly, and that coupling constants also do — that (colloquially) all have the proper units (sec. I.2.a.v, p. 12). But units are defined by the dilatation transformation; the unit of any object, relative to that of the coordinates (or mass), is determined by the requirement that probability densities and equations governing statefunctions all transform appropriately.

Invariance of the free Dirac equation again requires that mass and coordinates transform inversely under dilatations, that is dilatations relate units of mass to those of coordinates. And there are conditions on interactions when they appear. Also for both these equations to be invariant the masses of the objects must transform the same under dilatations, that is have the same units — so the ratio of masses is invariant (sec. I.2.a.vi, p. 13). This, not that masses be zero, is what is required for Dirac's equations to be invariant, provided interaction terms transform properly.

IV.3.d   Transformations of coupling constants under dilatations

How do coupling constants transform under dilatations, why, and what are their values? They should be invariant. Are they?

Dilatations place restrictions on interactions (sec. I.2.a.iv, p. 11). We first pick a set of units ($h = c = 1$, always and $m = 1$, where m is the mass of some object, say the electron), then coupling constants (except apparently the gravitational) are pure numbers. This is clearly true of the electron charge (but would not be if e, c, h, were in different systems of units, say one metric, the others using yards and inches respectively as units of length).

There are conditions on coupling constants for they appear in equations, like Dirac's equation, that have to be dilatationally invariant, determining their transformation properties under these transformations, more familiarly called their units. It is important that each appears in several equations of different form, and all give the same units (fortunately). Perhaps it should be obvious that they must. But it is still nice that they do.

IV.3.d.i   *The behavior of electric charge under dilatations*

Perhaps the easiest to analyze, and also the most familiar, is the electromagnetic coupling, which has an additional requirement, gauge invariance. The charge of the electron (from the fine-structure constant) is a pure number, it has no units. Why?

Consider, for example, Dirac's equation

$$i\gamma_\mu \frac{d\psi(x)}{dx_\mu} - M\psi + ie\psi\gamma_\mu A_\mu = 0. \qquad (\text{IV.3.d.i-1})$$

Now γ_μ is invariant under a dilatation,

$$d_\mu \Rightarrow \lambda d_\mu, \quad \text{so} \quad x \Rightarrow \frac{1}{\lambda}x, \quad M \Rightarrow \lambda M, \qquad (\text{IV.3.d.i-2})$$

as we expect, and

$$A_\mu \Rightarrow \lambda A_\mu. \qquad (\text{IV.3.d.i-3})$$

Invariance of Dirac's equation under dilatations, with e dimensionless, requires $A \Rightarrow \lambda A$. Then under a gauge transformation

$$\psi \Rightarrow \psi exp(ie\phi), \quad A_\mu \Rightarrow A_\mu - \frac{d\phi}{dx_\mu}, \qquad (\text{IV.3.d.i-4})$$

so ϕ and $e\phi$ are dimensionless thus e is.

The equation for A, the expression of the Poincaré Casimir invariant, is

$$\frac{d^2 A_\mu}{dx^2} = -ie\psi^* \gamma_\mu \psi, \quad \text{giving} \quad \psi \Rightarrow \lambda^{\frac{3}{2}} \psi, \qquad (\text{IV.3.d.i-5})$$

as we know. The ψ transforms the same whether this is determined using this equation (since e is dimensionless), or Dirac's equation — the requirements are consistent (fortunately). Gauge invariance gives

$$A'_\mu = A_\mu - \frac{d\phi}{dx_\mu} \Rightarrow \lambda A'_\mu = \lambda A_\mu - \lambda \frac{d\phi}{dx_\mu}, \qquad \text{(IV.3.d.i-6)}$$

as required for both equations to be invariant. So ϕ and the charge are both invariant under dilatations — they have no dimension. There are two requirements on these equations, invariance under both dilatations and gauge transformations. They can only be satisfied simultaneously if the charge is dimensionless, it has no units. And fortunately they can both be satisfied.

Charge density is proportional to probability density times the particle's charge which requires that charge be a pure number (times e). This is not true for strong interactions. There is no quantity corresponding to charge for nucleons, which allows more freedom for the strong interaction coupling constant.

The electromagnetic coupling constant, the charge, is

$$e = \sqrt{\alpha} \sim \sqrt{\frac{1}{137}} \sim 0.085. \qquad \text{(IV.3.d.i-7)}$$

IV.3.d.ii *The coupling constant for the weak interactions*

Under a dilatation the Fermi coupling constant $G_F \Rightarrow \lambda^{-2} G_F$. However this constant really does not exist being based on a nonexistent four-fermion interaction. It is phenomenological, and there is a mass involved, that of the intermediate boson, M_w. The physical coupling constant is that for the interaction of a fermion and an intermediate boson. This is approximately [Particle Data Group (2000), p. 95]

$$G_F = \frac{\sqrt{2}g_w^2}{8M_w^2}, \qquad \text{(IV.3.d.ii-1)}$$

so

$$G_F = 1.2 \times 10^{-5} Gev^{-2} = \frac{\sqrt{2}g_w^2}{8M_w^2} = \frac{\sqrt{2}g_w^2}{8(80Gev)^2}. \qquad \text{(IV.3.d.ii-2)}$$

Thus, roughly,

$$g_w^2 = \frac{8}{\sqrt{2}}(1.2 \times 10^{-5})(80)^2 = 0.4, \qquad \text{(IV.3.d.ii-3)}$$

$$g_w = 0.2, \qquad \text{(IV.3.d.ii-4)}$$

a pure number.

Why a pure number? A coupling constant with units would require an intrinsic mass or length in the theory. But since there is no length in geometry, there is no length to appear in the governing equations. And in the interaction between a boson (intermediate or photon) and a fermion there is no mass that can enter. There is one weakness in this argument, and it is interesting that the exception does not occur. The coupling constant might depend on the masses of the interacting objects, so the weak interaction would be different for say electrons and muons. But in fact the interaction, the coupling constant, is universal, independent of the object in whose equation the interaction appears, the same for all. Perhaps this should be noticed, for it might be significant.

Despite the interaction being weak, its coupling constant is actually larger than that of the electromagnetic interaction. It is the mass of the intermediate boson that results in the weakness of the weak interaction.

IV.3.d.iii *Coupling for strong interactions*

To see the difference between electromagnetism and interactions with other types of particles consider the Dirac equation for an interaction with a scalar π (parity is irrelevant here). The Klein-Gordon equation for π — expressing the Poincaré invariant — behaves the same as the one for A under dilatations (sec. IV.3.d.i, p. 206), so

$$\pi \Rightarrow \lambda \pi, \qquad\qquad\qquad \text{(IV.3.d.iii-1)}$$

as we know (eq. IV.3.c.i-6, p. 203). This also follows from the usual relationship between π and the number (or energy) operator [Schweber (1962), p. 160, 167]. Dirac's equation is

$$i\gamma_\mu \frac{d\psi(\lambda x)}{dx_\mu} - M\psi + g\pi\psi = 0. \qquad\qquad \text{(IV.3.d.iii-2)}$$

With

$$g\pi \Rightarrow g'\pi' = \lambda g'\pi, \qquad\qquad \text{(IV.3.d.iii-3)}$$

g is dimensionless, exhibiting the transformation for the strong coupling constant.

With derivative coupling in the equation for π from the Poincaré invariant,

$$(d_\mu^2 - m^2)\pi + ig\gamma_\mu d_\mu \psi\psi + ig\psi\gamma_\mu d_\mu\psi = 0, \qquad \text{(IV.3.d.iii-4)}$$

$$\lambda^3(d_\mu^2 - m^2)\pi + \lambda^4 ig'\gamma_\mu d_\mu\psi\psi + \lambda^4 ig'\psi\gamma_\mu d_\mu\psi = 0, \qquad \text{(IV.3.d.iii-5)}$$

requiring that this coupling constant have units, thus again arguing against derivative coupling.

Likewise interactions containing powers of boson statefunctions, $ig\psi\pi^2$ for example, can be ruled out. Requiring then that all bosons transform the same under dilatations, and that coupling constants be dimensionless, places strong restrictions on interactions (as does, in addition, invariance under the Poincaré group). Presumably they all do transform this way, and all interactions with them are of this form, which is of some interest if true. But for strong interactions it is impossible to investigate this experimentally, so we cannot be sure that it is true.

IV.3.e Transformation of the gravitational coupling constant and reasons why

To clarify why coupling constants are pure numbers, consider the gravitational constant, the one exception; also it is very small compared to the others (using any reasonable value for a unit mass). The gravitational coupling constant behaves under dilatations as $G \Rightarrow \lambda^{-2}G$, as does the Fermi constant. But for gravity the constant gives the coupling of a boson to fermions, as well as to other bosons. There is no indication that there is an intermediate boson for gravitation with the coupling constant determined by its mass (although there is a possibility that a Planck mass particle has never been discovered because it has never been, or could not have been, looked for).

Thus the gravitational constant is not a pure number because there is an intrinsic mass (although what this means for gravitation is not clear) and the units for the constant depend on the ratio of it to the unit mass.

Because gravitation is necessarily nonlinear the gravitational field (the connection) transforms under dilatations according to [Mirman (1995c), sec. 8.2.1, p. 146]

$$\Gamma \Rightarrow \lambda\Gamma, \tag{IV.3.e-1}$$

since the equation for its covariant derivative contains both the ordinary derivative and Γ^2. This is different from the transformation of the, say, electron field ψ (sec. IV.3.c.i, p. 202). The gravitational coupling constant is defined from the equation [Mirman (1995c), sec. 9.3.4, p. 161], schematically,

$$\Gamma^\beta_{\alpha\nu;\mu} = G\delta^\beta_\alpha T_{\mu\nu} + \dots, \tag{IV.3.e-2}$$

where T is the energy-momentum tensor. So [Mirman (1995c), sec. 9.2, p. 153]

$$\lambda^2\Gamma'^\beta_{\alpha\nu;\mu} = G'\lambda^4\delta^\beta_\alpha T'_{\mu\nu} + \dots, \tag{IV.3.e-3}$$

giving

$$G' = \lambda^{-2}G. \tag{IV.3.e-4}$$

Note that

$$G \sim \frac{1}{M_p^2} \Rightarrow G' = \frac{1}{M_p'^2} = \frac{1}{M_p^2}\lambda^{-2} = \lambda^{-2}G, \tag{IV.3.e-5}$$

with M_p the Planck mass.

IV.3.e.i *Differences between couplings*

There are several differences between the coupling of gravitation and
that of the electron to the electromagnetic field. In Dirac's equation
the electromagnetic coupling term contains the statefunction of the
electron, but the term for the interaction of gravitation with an object does not contain the gravitational statefunction (the connection).
As the interaction term for an electron with the electromagnetic field
incorporates the electron's statefunction, its strength depends on the
density of electrons. However this is misleading as the field interacts
with only one electron, although the interaction term does depend on
the charge distribution in space. But for gravitation the interaction term
(the energy-momentum tensor) is independent of the gravitational field.
The gravitational field produced by an object does not depend on the
gravitational field. This is also true for the electromagnetic field.

The second difference is that because of nonlinearity the gravitational statefunction has units of mass. Thus it cannot be regarded as
giving directly a probability density. However the concept of a probability for "gravitons" is highly questionable, since that of "gravitons"
itself is [Mirman (1995c), sec. 11.2.2, p. 187].

It is interesting that the required nonlinearity of gravitation determines its transformations under dilatations — its units.

Thus gravitation interacts very differently, and is quite different,
from other fields. It is not clear whether this is a clue to something.

IV.3.e.ii *How dilatations determine the behavior of free gravitation*

Gravitation is nonlinear, the equation for the free gravitational field
(the momentum operator acting on connection Γ) includes terms of
second-order in Γ [Mirman (1995c), sec. 8.2.1, p. 146]. Can there be
higher-order terms? The reason that second-order terms appear is that
gravitation must be nonlinear [Mirman (1995c), chap. 4, p. 56]. Why
are there no higher powers of Γ? The answer is dilatation invariance. It
gives the covariant derivative of the gravitational connection and shows
that higher-order terms cannot be present — this momentum operator
does not contain coupling constants, so the requirement is independent of how such constants might vary under dilatations. There is no
freedom in choice of units for constants, as there are no constants.

As $\Gamma \Rightarrow \lambda\Gamma$ under dilatations, $\Gamma_{,\mu}$ and $\Gamma\Gamma$ have the same units (that is
both transform the same way under dilatations), so in equations for the
covariant derivative (the momentum equations) they can be added, as
is necessary for the required nonlinearity. But $\Gamma\Gamma\Gamma$, or higher powers,
would have extra powers of λ — would not have the same units — so
these cannot appear. The equation for gravitation is nonlinear, containing (necessarily) terms of second-order in Γ, but only such terms.

This then completes the derivation of the equations governing free gravitation [Mirman (1995c), chap. 8, p. 135], showing that the expression obtained is correct, as no other terms can occur. Thus the behavior of free gravitation — and so to a large extent all gravitational fields — is determined (by its representation of the Poincaré, and similitude, groups).

IV.3.e.iii *The Gravitational Constant distinguishes objects*

What are implications of the gravitational constant G not being a pure number? One is that objects can be distinguished by their acceleration. Using Planck mass, M_p, with corresponding wavelength Λ_p, as the unit, $G = 1$. These are related by

$$G = \frac{1}{M_p^2} = \Lambda_p^2. \tag{IV.3.e.iii-1}$$

Two point masses, each m_1, are placed a distance of their Compton wavelength Λ_1 apart — that there are no such things as point masses is irrelevant here, as only formulas are considered; we can put them at $1000\Lambda_1$. They have to be put at some known multiple of this distance because there is no other distance known — there are no other objects, except gravitation, involved (ignoring irrelevant measuring instruments). The acceleration of one is measured (quickly before they move much so change the distance), and is

$$a = G\frac{m_1}{\Lambda_1^2} = \frac{1}{M_p^2}\frac{m_1}{\Lambda_1^2} = \frac{m_1^3}{M_p^2}, \tag{IV.3.e.iii-2}$$

so

$$a = m_1^3, \tag{IV.3.e.iii-3}$$

in terms of the unit mass,

$$M_p = 1. \tag{IV.3.e.iii-4}$$

Thus the measured value of acceleration a for an object depends on the ratio cubed of its mass to the square of the Planck mass, that is its mass cubed. This value is invariant under dilatations (acceleration is proportional to an inverse length),

$$m_1 \Rightarrow \Lambda m_1, \quad M_p \Rightarrow \Lambda M_p, \quad a \Rightarrow \Lambda a, \tag{IV.3.e.iii-5}$$

as it must be. Dilatations can have no experimental effect, so physical laws must be invariant under them.

This means that if we have two identical particles in space, far from other objects (ignoring measuring instruments), we could tell whether these are, say, electrons or muons (without waiting for the muon to decay). If we take the mass of the particles as 1 (there being no other objects to use for units), and their Compton wavelengths as the unit length (and unit time; $c = 1$), for there is nothing else, and measure the gravitational acceleration of one due to the other, then, even though everything seems identical, the acceleration, measured in these units, is different. So gravity gives a fundamental distinction between objects that are identical, except for different masses. The reason is that gravitation has an intrinsic mass, the Planck mass, and gravitational acceleration depends on the ratio of the mass of an object to this mass. And this ratio is different for the electron and muon.

IV.3.f Accelerations in terms of the mass levels

Masses of (charged) particles (appendix B, p. 246) lie close to mass levels (with the a term neglected)

$$m_n = \frac{nm_e}{\alpha}, \quad \Lambda_n = \frac{\alpha \Lambda_e}{n}, \tag{IV.3.f-1}$$

where m_e is the electron mass, Λ the Compton wavelength, α the fine-structure constant, and n is an integer or half-integer. For the muon

$$n_\mu = \frac{3}{2}. \tag{IV.3.f-2}$$

Thus

$$m_e = \frac{m_n \alpha}{n}, \quad \Lambda_e = \frac{n \Lambda_n}{\alpha}. \tag{IV.3.f-3}$$

Now the gravitational acceleration of an object with mass value m_n given by n is, in terms of the Planck mass M_p,

$$a_n = \frac{m_n^3}{M_p^2} = \frac{Gn^3 m_e^3}{\alpha^3}, \tag{IV.3.f-4}$$

which gives the gravitational acceleration in terms of the electromagnetic fine-structure constant.

IV.4 Differing Interactions Require Baryon and Lepton Conservation

Baryon and lepton numbers are conserved. Why? Baryon number must be because baryons are subject to strong interactions, leptons are not. Conservation of baryons leads to that of leptons. This raises further questions which are noted. While this is known [Mirman (1998)], it is important to discuss it using the formalism considered here.

A fundamental attribute of elementary particles is their set of conservation laws. Some arise from the nature of space [Mirman (1995a), sec. I.6.d, p. 27; sec. X.7.e, p. 299], but others are nongeometrical, such as conservation of baryons, and following from it, conservation of leptons. Conservation of these quantum numbers comes not from space but from their interactions.

This discussion is schematic, excluding everything unneeded, which simplifies the writing and aids understanding, and by not including the irrelevant, emphasizes the generality of results. We ignore electromagnetism and gravitation and their labels. They do not allow decay of protons. To be concrete, and as it is often discussed this way, we speak of the decay of the proton, but this applies to any baryon (a fermion with strong interactions) or antibaryon, and pions, but that can be any meson (a boson with strong interactions). Also lepton (a fermion with no strong interactions) means either lepton or antilepton. All three types of particles have weak interactions that would cause, if it were possible, the proton to decay into leptons plus pions, or directly into leptons (it does not matter if the decay is attributed to other interactions; the label weak merely denotes the interaction, whatever it is, whatever its properties, that causes the decay that we wish to show is impossible). Strong interactions are often stated as plural, for generality. As there are no other stable particles (known), the analysis indicates that if there are several types, every baryon is affected by each.

IV.4.a How interactions lead to conservation

Why is baryon number conserved? If it were not a particle with strong interactions would go into ones without; strong interactions would be "turned off". But weak interactions (or any other ones) cannot "turn off" strong ones (and thus cannot turn them on). There is no Hamiltonian with one interaction that changes another interaction. We first give an intuitive picture, then a formal argument. Purely heuristically, to see what goes wrong, take the unphysical case of the proton having weak interactions, the pion not. This gives a Feynman diagram in which the proton emits a virtual pion and then decays into three leptons. But the pion, being virtual, has to be reabsorbed. The leptons, however, do not interact with it, so it cannot be reabsorbed leaving it in a very unpleasant situation — implying that protons cannot so decay (and that conservation of energy would fail if they did). Similarly, a state only of leptons, with no strong interactions, cannot go into one with baryons (without an equal number of antibaryons). The reaction is not possible in either direction.

While the weak interaction of baryons cannot turn the strong one off or on, mesons do decay into leptons. Why? The photon is analogous; its number is not conserved, though charge is. However it is neutral, it couples to a neutral object, taken as a particle-antiparticle pair (the current is of this form). Electron-photon scattering can be regarded as the creation of a pair by the photon; the positron annihilates the electron, and the electron of the pair replaces it. So while the electromagnetic interaction (of an electron) cannot be altered, photons can be taken as not having a direct interaction (as indicated by the form of the current). Their creation or annihilation does not modify an interaction. It is similar for mesons, which can be deemed to have no strong interactions, but rather couple to particle-antiparticle pairs. We can view meson-baryon scattering as annihilation of the baryon by the antiparticle of the pair, and its replacement by the baryon of the pair. Decay can be considered as the creation of a pair and its annihilation into leptons or photons. Because of what they are coupled to, mesons can decay into objects not affected by strong interactions.

IV.4.b   Mathematical analysis

With this intuitive understanding of why baryons cannot decay into only leptons, but mesons can, we turn to a formal analysis, based on the formalism for quantum field theory developed above.

To study the decay we consider the action of the Hamiltonian, H, on a proton. Acting on a state at $t = 0$, time-translation operator $exp(iHt)$, gives the state at time t. Can this take a baryon to a state whose fermions are only those not having strong interactions? The effect of this is seen from that of H, a sum of the free particle Hamiltonians for the proton, pion and leptons, plus interaction terms for the weak and strong interactions. The state of the system is a sum of terms, one the state of the proton, another (if decay were possible) the product of pion and lepton states (plus products of lepton states if there were those decays), each sums over other labels and integrals over momenta or space. Take the initial state as a proton, say at rest. The free part of H changes its phase. The weak interaction part, were decay possible, decreases its coefficient in the sum, while increasing that of the (say) pion plus lepton, initially zero — starting as a proton, the state becomes a sum of the proton, its contribution decreasing, plus the pion plus lepton state, with increasing contribution. For the decaying pion the behavior is similar: starting as a pure pion it becomes a sum of that plus a state of leptons, with the contribution of the first decreasing, of the second increasing.

What is a proton? We define it as a particle obeying Dirac's equation, with mass m_P, where this equation includes the weak, strong and (irrelevant and suppressed) electromagnetic interactions (whose forms are not needed). It is the presence of these interactions that determines what a proton is. (Correctly a particle is an eigenstate of the two Poincaré invariants [Mirman (1995b), sec. 6.3, p. 114]. For a free spin-$\frac{1}{2}$ particle, and one with an electromagnetic interaction, Dirac's equation is equivalent. Whether this is true with weak and strong interactions seems unknown so consequences of, perhaps important, differences, if any, are not clear. And it is possible that putting interactions in invariants, which must be done whether Dirac's equation is used or not, might limit them. Particles are also eigenstates, or sums, of the momentum operators [Mirman (1995c), sec. 5.4, p. 93], of which the Hamiltonian is one. We ignore these, and refer to Dirac's equation as it is familiar, but the discussion could be of invariants, which might be revealing). The statefunction of the proton then is a solution to coupled nonlinear equations. We need information about it but cannot solve explicitly so represent it in a way that allows analysis, using an expansion. The arguments though are exact; we do not calculate, so need not truncate.

IV.4.b.i *Analysis using physical statefunctions*

A physical particle (sec. IV.2.d, p. 181), labeled with a capital, that obeying Dirac's equation with all interactions, is a sum of states (schematically):

$$|P) = c(x,t)|p) + \sum c(x,t)_{p\pi}|p)|\pi) + \ldots + \sum c(x,t)_{K\Lambda}|K)|\Lambda) + \ldots,$$

$$\text{(IV.4.b.i-1)}$$

summing over all states to which the proton is connected by interactions, including any number of pions, and so on. The summations represent ones over internal labels and integrals over momenta. These, and all other irrelevancies, like spin, are suppressed. That they can be shows the generality of this. State $|p)$ is the function satisfying Dirac's equation with the weak interaction, but not the strong — its effect is given by this sum, which is thus an eigenstate, with mass m_P, of the total Hamiltonian, including all interactions. Individual terms in the sum differ in energy and momenta; it is the sum that has the eigenvalues. We can take $|\pi)$ as either bare or physical; for the former each term is an infinite sum, which is encapsulated by regarding it as physical.

The coefficients are determined by the requirement that this be a solution of the complete Dirac equation, and normalization $(P|P) = 1$. Also the initial state, say a proton at rest, gives $c(x, 0)$, a wavepacket, with every other $c = 0$ at $t = 0$; they depend on the statefunction for $|P)$. These particles are virtual in not obeying the physical Dirac equation, that with interactions. Only P is physical. Dirac's equation (with interactions schematic, and higher-order terms not excluded) is

$$iy_\mu \frac{d|P)}{dx_\mu} + g|\pi)|P) - m_P|P) = 0$$

$$= iy_\mu \frac{dc(x,t)}{dx_\mu}|p) + g|\pi)|p) - m_P|p) + iy_\mu \sum \frac{dc_{p\pi}(x,t)}{dx_\mu}|p)|\pi)$$

$$+ g|\pi)^2|p) - m_P|p)|\pi) + \ldots = 0. \quad \text{(IV.4.b.i-2)}$$

States of different numbers of pions are orthogonal, so this gives an infinite set of coupled equations for the c's; solving (in principle) gives the state of the (physical) proton. (We need not consider how far this can be taken to find, and solve, recursion relations to obtain physical states; doing so for a particle that does decay might provide information about it, and its other interactions.) This expansion has an infinite number of pions, that is terms $|p\rangle|\pi\rangle^r$, for all $r \geq 0$, because (writing $|p\rangle$ and $|\pi\rangle$ using creation and annihilation operators acting on the vacuum) the strong part of the interaction Hamiltonian is (schematically)

$$H_s \sim \sum_r a^* a (b + b^*)^r; \qquad \text{(IV.4.b.i-3)}$$

a is the annihilation, a^* the creation, operators for $|p\rangle$, the $b's$ the ones for pions. This acting on $|p\rangle|\pi\rangle^r$ gives $|p\rangle|\pi\rangle^{r+1}$ and $|p\rangle|\pi\rangle^{r-1}$, thus the expansion includes all $r \geq 0$.

Now the weak interaction acts on $|p\rangle$ supposedly causing it to decay, so the final state is

$$|fs\rangle = \sum d_{\pi l}|\pi\rangle|l\rangle + \sum e_{\pi l} \sum c_{p\pi}|l\rangle|\pi\rangle|\pi\rangle + \dots$$
$$+ \sum d_{lll}c|l\rangle|l\rangle|l\rangle + \sum d_{\pi lll} \sum c_{p\pi}|l\rangle|l\rangle|l\rangle|\pi\rangle + \dots, \quad \text{(IV.4.b.i-4)}$$

showing the transition to, say, a pion plus a lepton, and to three leptons, with coefficients of non-occurring terms zero. The energy of $|fs\rangle$ is m_P, not (the unphysical) m_p, so needs contributions from $\sum c_{p\pi}|l\rangle|\pi\rangle|\pi\rangle$, and so on. However $|fs\rangle$ is, say, a lepton plus a pion, so these other terms, to which this is orthogonal, cannot contribute. The decay of the proton cannot conserve energy, thus cannot occur. That is the supposed decay is of the unphysical particle, with no strong interactions, but the proton is the physical particle, with these interactions.

It is not possible to have the $|p\rangle$ in each term of the expansion decay because that would give terms with any number of pions, plus other particles to which the proton, or pion, or other particles, were coupled. There are also terms in the expansion not having a $|p\rangle$. And the energy of these would be arbitrarily greater than that of the physical proton.

This can also be considered by using a Hamiltonian that is a sum of three terms, a free part with no interactions, a weak part causing the supposed decay and a strong part. The sum is the operator that acts on the physical particle. Thus it (presumably) takes a physical proton to a sum of itself plus $|decay products\rangle$, schematically

$$H|P\rangle = (H_f + H_w + H_s)|P\rangle \Rightarrow c_p(t)|P\rangle + c_{dp}(t)|decay products\rangle$$
$$= c_p(t)|p\rangle + \dots, \quad \text{(IV.4.b.i-5)}$$

using the expansion for $|P\rangle$. If the strong interaction is not considered then the only term in the expansion is $|p\rangle$ — this is the solution of the equation with only the free and weak parts. Hence it is the weak part of the Hamiltonian that causes the decay, that causes this (nonphysical) part of the proton's statefunction to evolve to one describing a sum of the (nonphysical) proton plus decay products. But as we see this does not allow the physical proton to decay.

Likewise decays of leptons (the τ) to baryons are similarly ruled out.

IV.4.b.ii *Decay of the* Δ

Contrast this with the strong decay of the Δ; the physical one is Δ_p, while Δ satisfies Dirac's equation without the strong interaction (thus with no interaction); π_p is the physical pion, π satisfies interaction-free equations. Then Δ_p has an expansion similar to the proton's, except that it can decay by the strong interaction into a proton plus a pion giving extra terms (with coefficients w),

$$|\Delta_p) = h(x,t)|\Delta) + \sum h(x,t)_{\Delta\pi}|\Delta)|\pi) + \ldots + w(x,t)|\pi_p)|P)$$

$$= h(x,t)|\Delta) + \sum h(x,t)_{\Delta\pi}|\Delta)|\pi) + \ldots + w(x,t)c(x,t)|p)|\pi_p)$$

$$+ \sum c(x,t)_{p\pi}|p)|\pi)|\pi_p) + \ldots . \quad \text{(IV.4.b.ii–1)}$$

The $|\Delta)$, having no interactions, does not decay, so this expansion cannot be used for an argument like that for the proton's weak decay — $|p)$ and $|\Delta)$ are solutions of the free-particle Dirac's equation, there is only a single interaction the strong one (which causes the decay), so the argument fails. However the physical $|\pi_p)|P)$, for example, can have the same energy as the $|\Delta_p)$. This decay, of the entire sum, is possible. (But $|\Delta)$ cannot decay into particles with no strong interaction.)

IV.4.b.iii *Why the neutron can decay*

The neutron has an actual weak decay; why do the arguments not apply? Taking it at $t = 0$ in a single particle state gives

$$|N) = c(x,0)|n) + \sum c(x,0)_{n\pi}|n)|\pi) + \ldots, \quad \text{(IV.4.b.iii–1)}$$

where $|n)$ is the particle obeying Dirac's equation with the weak interaction, but not the strong, and $|N)$ is the physical particle, with both interactions. The $|n)$ decays to $|p)|l)|l)'$, so this becomes

$$|N) = c(x,t)|n) + \sum c(x,t)_{n\pi}|n)|\pi) + \ldots + \sum d(x,t)|p)|l)|l)'$$

$$+ \sum d(x,t)_{p\pi}|p)|\pi)|l)|l)' + \ldots . \quad \text{(IV.4.b.iii–2)}$$

The d-coefficients give the physical state $|P)|l)|l)'$. The state then is

$$|fs) = w_N(t)|N) + w_{p2l}(t)|P)|l)|l)'; \quad \text{(IV.4.b.iii–3)}$$

the first coefficient decreases in time, the second increases, so the total probability is constant. These two (orthogonal) states have equal energy. The neutron can decay into a strongly-interacting particle (the proton), but not otherwise, because it is state $|n\rangle$ that decays to $|p\rangle$. But the physical neutron is a sum of terms like $|n\rangle|\pi\rangle^r$, so the resultant proton is a sum of such terms as $|p\rangle|\pi\rangle^r$, which is the physical proton state — the proton, affected by the strong interaction, is such a sum (plus others appearing also for the neutron). Thus a physical particle decays into a physical particle. If the neutron were to decay into state $|L\rangle$ without the strong interaction, the expansion would have terms like $|L\rangle|\pi\rangle^r$, which do not sum to a state of any physical particle — there is no final state with this expansion so the matrix element of the Hamiltonian causing such decay is zero; decays annulling the strong interaction cannot occur.

IV.4.c   Thus leptons are conserved

Conservation of baryons implies that of leptons. The rotation group requires that the numbers of fermions in the states before and after a reaction or decay must both be odd, or both even, and we assume that the number of particles minus antiparticles is constant (it is not clear that otherwise a hermitian Hamiltonian is possible especially one that does not lead to difficulties like those above, especially if neutrinos have nonzero mass, as we expect [Mirman (1995c), sec. 4.4.4, p. 70], which would rule out antineutrinos being the same as neutrinos). Then leptons cannot be produced or destroyed if baryon (minus antibaryon) number is constant — lepton number is conserved.

The pion can decay into leptons: Why does the argument not apply? The physical pion $|\pi_p\rangle$ can be written

$$|\pi_p\rangle = c|\pi\rangle + c_{PP'}|PP'\rangle; \qquad (IV.4.c-1)$$

$|\pi\rangle$ obeys Dirac's equation with strong interactions, but not the weak — this causes the decay — and the prime indicates antiparticles. This contains two terms (plus irrelevant ones for other particles coupled to the pion). The strong part of the pion's Hamiltonian

$$H_S \sim A^*A^{*\prime}B + AA'B^*, \qquad (IV.4.c-2)$$

acting on a pion gives a particle-antiparticle pair, acting on this pair gives the pion, and similarly for r pions which is mixed with the pair plus $r - 1$ pions. Only two states appear in the expansion for the pion, unlike the infinite number for the proton (we ignore such terms as ones due to more than a single pair, for example). The weak part of H acting on $|PP'\rangle$ causes it to decay to two leptons; this is related by crossing to proton-antiproton scattering into two leptons, and also to the decay $N \Rightarrow P + 2l$, which we saw is possible. Thus both terms in the sum for the physical particle contribute to the decay, it does conserve energy (and momentum), therefore cannot be ruled out.

The argument for electric charge conservation is the same. Charge conservation is related to gauge invariance, a partial statement of Poincaré invariance [Mirman (1995c), sec. 3.4, p. 43], relating an allowed interaction to the Poincaré group. An interaction violating charge conservation would not transform under gauge transformations as other terms in the Hamiltonian, giving Poincaré transformations (on massive objects) that induce gauge transformations (on massless ones) resulting in physically-identical observers who undergo the different gauge transformations — these cannot be fully specified — thus physically identical, but who see different Hamiltonians. The Hamiltonian would not be well-defined, implying inconsistent physics. It is fortunate that charge is conserved.

IV.4.d   Why the argument for gravitation is different from that for conservation of baryons

Gravitation raises another point. There are reasons why it may be unable to act on scalar particles: pions, kaons, ... [Mirman (1995c), sec. 4.2.7, p. 61]. Assuming that it acts on vectors, like ρ, then objects affected by gravity decay to ones that are not (the ρ to pions for example), and ones not affected go to ones that are (such as pions to leptons). Suppose that there were particles that are unaffected by gravitation, as we expect ones with zero spin (scalars) to be [Mirman (1995c), sec. 9.2.2, p. 155]. Could they decay into ones on which gravity acts? So could the pion (if it had no gravitational interaction) decay into leptons, could the kaon (for which experimental checks are more plausible)? Could a $\tau\tau'$ annihilate into these scalar particles?

Does the argument not forbid this? It fails because there is no such thing as a state of, say, two "gravitons" [Mirman (1995c), sec. 11.2.2, p. 187], nor even a free "graviton", since gravity is — necessarily — nonlinear [Mirman (1995c), sec. 4.4.3, p. 69]. The expansion (sec. IV.2.d.ii, p. 182) therefore cannot be used for the gravitational interaction.

To see that the argument for conservation of baryons does not hold for these, consider the pion decaying into e and μ, so the state of the system is

$$|s) = c_\pi(t)|\pi) + c_l(t)|e)|\mu), \qquad \text{(IV.4.d-1)}$$

where at $t = 0$ the first is 1, the second 0, and the first decreases, the second increases. Now the $|e)$ and $|\mu)$ are physical states, ones obeying Dirac's equation with the gravitational interaction, which we must use since there are no expansions for them. The masses are those found experimentally, so include gravity. Hence the energy of the two terms is the same, and the decay conserves energy. Thus the argument that the proton cannot decay does not apply. This is similar to the argument showing pions can decay into leptons; that pions have strong interactions, leptons do not, is irrelevant.

Why does this argument for gravitation not apply to the weak decay of the τ? The reason is that it would decay into the unphysical proton because that is the object subject to the weak interaction that causes the supposed decay. But it is the physical proton that exists. If the τ could decay into the unphysical one it would decay into each of these states in the expansion of the physical object, and that would not conserve energy.

Thus the argument does not hold for gravitation, but we cannot argue the converse. We do not know theoretically whether such decays are possible. And from experimentally observed decays we cannot tell if scalar particles are affected by gravity or not — but clearly determining, experimentally (if possible), whether they are or not would be quite interesting. These questions remain open. Except for a (very) few particles, it is not even known (experimentally) which have interactions with gravity, nor what these are — something of great interest. Theoretical prejudice should not substitute for analysis and experiment. This emphasizes the difference between gravity and other interactions, how little is known, and the value of these questions as a probe into the laws of nature.

IV.4.e   Questions and implications

There are many implications requiring investigation; we mention a few in hope of stimulating such.

All interactions known are of lowest order. Why? For electromagnetism linearity is enforced by gauge (Poincaré) invariance [Mirman (1995c), sec. 4.2, p. 57]. For strong interactions, take a particle, a Δ or p, that emits a pion. Higher-order terms would couple it not to a single pion, but to more. Intuitively we can guess why only lowest order occurs since it gives diagrams which we interpret (purely heuristically) as two, or more, pions emitted sequentially. Higher-order means that these are emitted together. However this is the limit of the lowest order in which intuitively the time between emissions goes to zero. A higher-order interaction would be this limit, which is included in the lowest order as one case; higher-order terms adding nothing, would be irrelevant. Summing all diagrams, and integrating over time, would give contributions from terms that have the same effect as higher-order ones, thus changing only the value of the sum, so the value of the coupling constant — an experimental parameter (at present), thus we could not distinguish contributions from terms of different order, implying higher order would be undetectable. This regards particles as virtual. But consider a decay in two steps, each emitting a pion. If the intermediate object's life were sufficiently short, this would be equivalent to pions being emitted simultaneously. If a nucleon had an interaction of the form $NN'\pi$, the emission of an NN' pair could be thought of as due to the decay of a pion, and the interaction taken as the limit of the emission of a pion, and then its decay, when its lifetime becomes zero, merely changing the sum.

Suppose that the only particles were nucleons and pions, no excited ones. With a linear interaction final states of pion-nucleon scattering have only a single pion — the nucleon could not store the extra energy and decay to a second pion. Nonlinear interactions give states with more than one pion, but could be simulated by short-lived excited states — the pion scatters and excites the nucleon which then decays giving a second pion. Nonlinear interactions would be indistinguishable from existence of excited states; perhaps we could require all interactions be linear with addition of excited states. But these questions require greater rigor.

Why is the gravitational interaction of lowest order in the source term? The interaction of matter with gravity is clear [Mirman (1995c), chap. 5, p. 73]. But what determines how matter fixes the gravitational field [Mirman (1995c), chap. 9, p. 150]? The tensor could be $T_{\mu\nu}T_{\nu\rho}$, or could it? These are some questions that should be looked at.

These comments are merely suggestive, but rigorous discussions might be possible.

IV.4.f   Physics is constrained, and laws have reasons

There are strong constraints on physics as seen in many ways [Mirman (1995a), sec. I.2, p. 2; 1995b,c, 2001]. It is well known that interactions are greatly limited by the properties of space, but also because of other interactions. Perhaps this analysis, and more the questions it leads to, can induce further inquiry. There are reasons for the laws of nature, and it is fortunate that reasons, and laws, are such that we are able to find, and understand, them.

IV.5   Symmetry, transformations and physics

Physics and geometry are intimately related. What our physics is, what our geometry is, are each determined by the other, certainly greatly restricted by each other, and likely the properties of each are definitive in establishing those of the other. This has long been known, and emphasized often both here and before [Mirman (1995a,b,c;1999;2001)], and is quite widely known, although perhaps not as well-understood as it should be. Here we have tried to increase the impact of this understanding, so to provoke thinking that is so badly needed to obtain further understanding. How then have these views been strengthened by this discussion?

Transformation groups of geometry, the Poincaré group, and its subgroups, are quite familiar, as are requirements (although perhaps not all) that they impose on physics. But there is another group, of which the Poincaré group is a subgroup, that is closely related to geometry, the conformal group.

This group is unusual (or perhaps just not familiar) and complicated. Starting to explore it and its consequences, as has been done here, shows how it provides, and can provide, insight into physics (and geometry). And it shows how much remains to be done, that here we have taken just the smallest first steps. Yet this has allowed us, even if only very sightly, to clarify quantum mechanics, quantum field theory, how to formulate, how to use, them, and we have begun to see some further steps that should be taken, especially to elucidate the conformal group and what it can tell us about physics and geometry.

To construct a theory we use a set of states that are taken into each other by the operators of the group. Momentum (exponentiated) takes a state into a translated one, while a rotation operator (exponentiated) gives the rotated state (including boosted states in an indefinite metric space). Dilatations give states in which units are changed. The internal parts of the operators perform the corresponding transformations on the internal parts of the states (rotating spin for example). The theory then consists of all states produced by these operators. To construct a conformal field theory we enlarge the set of states to include all those obtained from the states of the theory by all conformal transformations (sec. IV.2.b, p. 171), including the internal parts of the states produced by these.

Here the method for obtaining the states given by the transversions has been outlined. They are obtained from the Poincaré basis states with momentum operators diagonal, so are different from ones usually found for semisimple groups. But these are the relevant states. Thus with this formalism as a foundation, further exploration of the physical and mathematical implications and applications can be started, including individual cases. Mathematical complications appear formidable, in part due to the novelty of the nature of the states and the nonlinearity of transversion operators. Thus it seems best, having set up the formalism, to leave applications to the detailed analyses for particular physical models — and to hope that others will join in this vast, but potentially highly rewarding, quest.

Note: An intuitive (and potentially misleading) statement is that its strong interaction gives a contribution to the proton's mass which disappears when it decays.

Appendix A

Lorentz group representations and their meaning

A.1   The Lorentz group as a subgroup

Subgroups are often essential in the development of representations and in interpretations of groups. The conformal group has a subgroup, the Poincaré group, which as the group of transformations of geometry is foundational. Its maximal homogeneous subgroup, thus the maximal homogeneous group of our geometry, is the Lorentz group. Beyond the importance it thereby obtains it is a noncompact group, with maximal compact subgroup the rotation group, so is interesting not only because of its geometrical and physical significance, but also because of lessons it may teach. Here we summarize some aspects that help provide insight about other topics that the book explores, and that may also suggest further subjects for exploration.

We have reviewed $SO(2,1)$, a subgroup of the Lorentz group, $SO(3,1)$ (sec. III.3.b, p. 133). Regarding it as a subgroup allows representations that cannot occur when it is taken as a group by itself. These are found by realizing states not on the space generated by the operators of $SO(2,1)$, but on the larger space of the Lorentz group. Of course the Lorentz group is itself a subgroup of bigger groups, so it and its subgroups can be realized on even larger spaces. This suggests various interesting possibilities, but here we can do no more than give an example of representations of the algebra of a group realized on a representation space of the algebra of a larger group of which it is a subgroup, that of $SO(2,1)$ the subgroup of $SO(3,1)$. Considering the group, not merely the algebra, may reveal further richness.

This is another aspect of the Lorentz group representations: they allow such representations of subgroups. Such freedom with group representations does not seem to have been much investigated, although perhaps it should be.

One present purpose is to emphasize this additional opportunity in conceivable realizations and corresponding representations, something that may not always be fully appreciated, but which might have important consequences and applications.

A.2 The meaning of Lorentz representation labels

Lorentz group representations are found from its little group [Mirman (1999), sec. VI.2.c, p. 284], the rotation group, and each Lorentz representation consists of a set of rotation representations. Representations of the Lorentz group are labeled by eigenvalues of two invariants [Gel'fand, Minlos and Shapiro (1963), p. 188; Naimark (1964), p. 116]. One gives the rotation representation with the smallest "angular-momentum" in the Lorentz representation — the smallest value of the label, the other, for a finite-dimensional Lorentz representation, gives the largest "angular-momentum". For an infinite-dimensional representation, eigenvalues of the second invariant determine matrix elements of the three other Lorentz operators (boost generators) — which are related because they are basis states of the rotation defining-representation; these change the rotation representation. The two eigenvalues give the coefficients of states of the three rotation representations obtained from a state on which a boost generator acts.

These infinite-dimensional representations we refer to as discrete representations. We also consider another type in terms of plane waves (sec. A.5.e, p. 239).

A.2.a   Why, physically, is a second label necessary?

What freedom is there, physically, that requires a second number to specify Lorentz representations? And why is there only one additional number — why if that is given are all algebra matrix elements determined?

Physically if an object, say an atom with angular momentum S in its rest frame, is put into a moving frame — boosted — its angular momentum state becomes a sum over states from an infinite number of irreducible rotation-representations. Statefunctions are products of internal states, those of spin, and space-dependent ones, plane waves. A plane wave can be expanded in a sum (of an infinite number) of spherical harmonics (sec. A.6, p. 240). The coefficients in this sum depend on the velocity, which is irrelevant here since we are considering the algebra not the group, and also the position of the origin with respect to the object. The angular momentum of an object moving on a line, so with a finite velocity, is zero when measured with respect to a point on the line,

$$\underline{L} = \underline{v} \times \underline{r} = 0, \quad \underline{r} \text{ parallel to } \underline{v}, \qquad \text{(A.2.a-1)}$$

but is a sum of different angular momentum values when measured with respect to a point off the line, and the coefficients in the sum depend on the distance between the line and the point (and velocity). If the state is a plane wave, the velocity is specified but the distance to the origin is not. And a wavepacket contains a set of velocities. While the velocity is determined by the group transformation (the boost), it does not depend on the position of the origin, so this must appear in algebra matrix elements.

A.2.b   Translation gives an infinite number of spherical harmonics

A translation of a spherical harmonic gives a function that is an infinite sum of spherical harmonics. Start with spherical harmonic $Y(\theta, \phi)^l_m$ [Mirman (1995a), sec. XI.4.c, p. 322; Varshalovich, Moskalev and Khersonskii (1988), p. 133], and the formulas relating θ, ϕ, to rectangular coordinates x, y, z [Mirman (1995a), eq. X.4.a.ii-2, p. 279]. A shift of the origin from $(0, 0, 0)$, to (u, v, w), transforms the angles [Varshalovich, Moskalev and Khersonskii (1988), p. 142], so, using the variables defined there,

$$\theta = \theta(\theta', u, v, w), \quad \phi = \phi(\phi', u, v, w). \qquad \text{(A.2.b-1)}$$

With

$$r^2 = x^2 + y^2 + z^2, \qquad \text{(A.2.b-2)}$$

and

$$\underline{r}' = \underline{r} - \underline{d}, \qquad \text{(A.2.b-3)}$$

$$r'^2 = r^2 + d^2 - 2rd\cos(\omega), \qquad \text{(A.2.b-4)}$$

$$\cos(\omega) = \cos(\theta)\cos(\zeta) + \sin(\theta)\sin(\zeta)\cos(\phi - \tau), \qquad \text{(A.2.b-5)}$$

$$\cos(\theta') = \frac{r\cos(\theta) - d\cos(\zeta)}{r'}, \qquad \text{(A.2.b-6)}$$

$$\tan(\phi') = \frac{r\sin(\theta)\sin(\phi) - d\sin(\zeta)\sin(\tau)}{r\sin(\theta)\cos(\phi) - d\sin(\zeta)\cos(\tau)}, \qquad \text{(A.2.b-7)}$$

with ζ, τ determined by the direction cosines of the displacement. Thus $Y(\theta, \phi)^l_m$ is expressed in terms of θ', ϕ', u, v, w, and can be written as expansions in terms of the Y's for θ', ϕ'. These are angular momentum states in the new coordinate system.

This can be illustrated using reducible representations of SO(3), homogeneous polynomials in x, y, z,

$$|l, \mu, \nu) = \sum \alpha(\mu, \nu) \frac{z^{l-\mu-\nu} x^\mu y^\nu}{r^l}, \tag{A.2.b-8}$$

summed over all values of positive μ, ν, with $0 \leq l - \mu - \nu$, where $\alpha(\mu, \nu)$ are coefficients chosen so that $|l, \mu, \nu)$ is an eigenfunction of L_z, and division by r is necessary to make these functions of angles, that is unit vectors.

For $l = 1$

$$|1, m = 0) = \frac{z}{r}, \quad |1, m = 1) = \frac{x + iy}{2r}, \quad |1, m = -1) = \frac{x - iy}{2r}, \tag{A.2.b-9}$$

and a translation gives

$$|1, m = 0)' = \frac{z + w}{2r'}, \tag{A.2.b-10}$$

$$|1, m = 1)' = \frac{(x + iy) + (u + iv)}{2r'}, \tag{A.2.b-11}$$

$$|1, m = -1)' = \frac{(x - iy) + (u - iv)}{2r'}, \tag{A.2.b-12}$$

$$r'^2 = z^2 + w^2 + 2zw + x^2 + y^2 + u^2 + v^2 + 2xu + 2yv. \tag{A.2.b-13}$$

The $l = 2$ irreducible representation has the unnormalized equivalent states,

$$|2, 2) = \frac{(x + iy)^2 - r^2}{r^2}, \tag{A.2.b-14}$$

$$|2, 1) = \frac{z(x + iy)}{r^2}, \tag{A.2.b-15}$$

$$|2, 0) = \frac{z^2 - r^2}{r^2}, \tag{A.2.b-16}$$

$$|2, -1) = \frac{z(x - iy)}{r^2}, \tag{A.2.b-17}$$

$$|2, -2) = \frac{(x - iy)^2 - r^2}{r^2}. \tag{A.2.b-18}$$

Translating results in

$$|2, 0)' = \frac{z^2 + w^2 + 2zw - r'^2}{r'^2}, \tag{A.2.b-19}$$

$$|2, 2)' = \frac{(x + iy)^2 + (u + iv)^2 + 2(x + iy)(u + iv) - r'^2}{r'^2}, \tag{A.2.b-20}$$

. . . .

First ignoring the change of r we see that the $|2,2\rangle'$ is a sum of the $(l = 2, m = 2)$, $(l = 1, m = 1)$, $(l = 0, m = 0)$, representations and states, as seen from the powers of $(x + iy)$ — referred to the displaced origin it is reducible, and these are the states in its decomposition. It is not invariant under a translation unless $u = v = 0$; however the $|2,0\rangle'$ state still has $m = 0$, but is a sum of the $l = 2$, 1 and 0 representations.

Clearly, equivalent results hold for all larger l. Thus translating the origin takes an angular momentum representation into a sum of representations, and also changes the μ values of the states, except when $u = v = 0$.

Under translation $r \Rightarrow r'$, and the angular momentum states are functions of $(r')^{-l}$, where l depends on the representation. However the coordinates are x, y, z, r, and Y has to be written in terms of these. Thus we must expand $(r')^{-l}$ in a power series, an infinite number of terms. Each angular momentum state then goes to an infinite sum of all angular momentum states, as expected physically.

A.2.c Arbitrariness of the origin requires second eigenvalue

Thus, because of this arbitrariness of the origin, to specify algebra matrix elements we must give one other number, and this is determined by the second invariant of the Lorentz algebra. Given the angular momentum in the rest frame and this distance (and velocity), the statefunction of an object — its Lorentz representation and state — is completely determined, so these two numbers, fixed by the two invariants, are necessary and sufficient. However while we can relate these eigenvalues to the physical state of the object, we do not need this picture; matrix elements are properties of the algebra, no matter how interpreted, and as this picture shows they are functions of two eigenvalues, so that is always true whether we use pictures or not.

Another way of seeing why two eigenvalues are needed is to consider the path of an object with orbital angular momentum, and also nonzero velocity, both parallel to z, that is projected onto the xy plane; its angular momentum is measured with respect to the point of intersection of the plane and the z axis. Thus to specify the statefunction we need the point of intersection of the velocity and the plane — there is a degree of freedom — and that is what the eigenvalue of the boost operator determines.

A.2.d   How translations and boosts give matrix elements

To illustrate determination of matrix elements consider translation d, and boost vt, both in the xz plane. Then for the Y_{10} spherical harmonic

$$Y_{10} = \frac{z}{r} \Rightarrow \frac{z + d_z + v_z t}{\sqrt{(z + d_z + v_z t)^2 + (x + d_x + v_x t)^2 + y^2}}, \qquad \text{(A.2.d-1)}$$

and to find the effect of algebra operator B_z, the component of the boost, we expand in terms of $v_z t$, take the first order term, and set all components $v_i t$ to 0. The result is a function of x, y, z and d. This is then expanded in terms of functions of x, y, z (spherical harmonics) whose coefficients are functions of d. Hence the coefficients of the algebra operators for boosts, giving the matrix elements, depend on where the origin is. And of course this is true for all spherical harmonics, boosts, and displacements.

So, with a Taylor expansion in $v_x t$, setting the velocity to 0,

$$\frac{z}{r} \Rightarrow \frac{z + d_z + v_z t}{\sqrt{(z + d_z + v_z t)^2 + (x + d_x + v_x t)^2 + y^2}}$$

$$\Rightarrow \frac{z + d_z}{\sqrt{(z + d_z)^2 + (x + d_x)^2 + y^2}} - \frac{(z + d_z)(x + d_x)}{\{(z + d_z)^2 + (x + d_x)^2 + y^2\}^{3/2}},$$

$$\text{(A.2.d-2)}$$

which is equal to a sum of an infinite number of spherical harmonics, with coefficients that depend on d.

It may seem that matrix elements depend on three parameters, the three components of the translation, rather than one. However in the limit as the velocity goes to 0 there is no way of distinguishing axes. Thus matrix elements can only be functions of the magnitude of a translation. The z axis can be defined to be along the translation, and the boost along z is then a function of only one parameter. As boosts form a vector under the rotation group knowledge of any one gives the others. If now axes are rotated, the translation and z axis no longer coincide, and the translation is given by three parameters. However these are the magnitude of the translation, and the angles of rotation. Matrix elements in the first system are functions of one parameter. The spherical harmonics whose coefficients they are under a rotation go into sums with coefficients that are functions of the rotation angles. Combining coefficients of the rotation and the boost gives the matrix elements of the boost in the transformed system, the coefficients of the spherical harmonics in this system. While these depend on three parameters, only one comes from the boost.

By the Wigner-Eckart theorem [Mirman (1995a), sec. XII.3.b, p. 350] matrix elements can be written as products of terms that are the same for all states of a representation, times terms that are state dependent.

A Lorentz state is a sum of spherical harmonics, with coefficients that are products, one term of which is the same for all states of a rotation representation, functions of boost matrix elements, these containing the magnitude of the translation, so are dependent on only one parameter. The second terms in the products are given by rotation matrix elements, not matrix elements of boosts, and these are functions of two parameters, the angles.

This algebra operator is independent of velocity, thus the state looks the same for all frames moving parallel to z; coefficients of various angular-momentum states, the fractions of the total state each gives, are the same for all frames — the state is invariant under a boost (a group, not an algebra, operator). It leaves the functional form of the statefunction the same, but changes the value of the variable on which the statefunction depends, the velocity, as a rotation changes the value of angle ϕ in $exp(im\phi)$, but does not change the function, specifically does not change m.

A.3   Types of eigenfunctions of representation-labeling operators

We wish to find the two commuting generators of the Lorentz group and their eigenfunctions which are needed to find and understand the group representations. While eigenfunctions of operator L_z, the spherical harmonics, are familiar, what are those of boost L_{zt}, and which are eigenfunctions of both? Actually boost generators, giving pseudo-rotations, have different types of eigenfunctions increasing the interest in them.

A.3.a   The commuting Lorentz labeling operators

One state-labeling operator, the generator of a rotation around z ($L_{xy} = L_z$), gives the weight of the state of the rotation representation [Mirman (1995a), sec. XIV.3, p. 408], its angular-momentum "z"-component, this determined by the value and orientation of the angular momentum in the rest frame. The Lorentz generator that it commutes with is that for a boost along z (L_{zt}). States are (taken as) eigenstates of these two operators.

That they commute be seen from the expressions for the two operators. For a boost along z the algebra operator is

$$L_{zt} = B_z = -i(z\frac{d}{dt} + t\frac{d}{dz}), \qquad \text{(A.3.a-1)}$$

which of course commutes with

$$L_z = -i(x\frac{d}{dy} - y\frac{d}{dx}).$$ (A.3.a-2)

A boost generator produces a sum of angular-momentum states, for L_{zt} all with the same z component. What physical state is invariant under this boost? It is one for which the origin is on the line of the velocity (properly along the line defined by the generator), the z axis.

To clarify the notation for boost transformation $exp(iuL_{zt})$ we consider an object with speed v in some system boosted with parameter u. Its speed then is [Rindler (1960), p. 34]

$$V = \frac{v + u}{1 + uv}.$$ (A.3.a-3)

The value of V varies from v, for $u = 0$, to

$$V_{max} = \frac{v + 1}{1 + v} = 1,$$ (A.3.a-4)

for a boost to the frame for which $u = 1$, that is on the light cone.

A.3.b   Eigenfunctions of exponential type

First, for this differential realization of L_{zt}, the eigenfunction is, with v the velocity,

$$|\beta\rangle = exp(i\beta arc\,tanh(\frac{z}{t})) = exp(i\beta arc\,tanh(v)).$$ (A.3.b-1)

This is also an eigenfunction of $L_z = L_{xy}$ with eigenvalue 0. It can be checked using the derivative

$$\frac{d(arc\,tanh(u))}{dx} = \frac{1}{1 - u^2}\frac{du}{dx} = \frac{d(arc\,cotanh(u))}{dx}.$$ (A.3.b-2)

If the coordinates are written, with the value of ρ irrelevant,

$$z = \rho cosh(\xi), \quad t = \rho sinh(\xi),$$ (A.3.b-3)

then

$$L_{zt} = -2i\frac{d}{d\xi}.$$ (A.3.b-4)

For this form the eigenfunction is

$$|\beta\rangle = exp(-i\beta\xi),$$ (A.3.b-5)

which is equivalent to $exp(im\phi)$ for a rotation but since ϕ is periodic m is integral while there is no such condition on ξ so β takes on all real values.

What is the significance of eigenvalue β? The eigenfunction is a function of velocity v — a function of the ratio $\frac{v}{1}$ (whose maximum value is 1, for the light cone). Thus β gives the rate, with respect to the change of velocity, at which the eigenfunction is changing (properly the rate of change of the eigenfunction with respect to ξ).

Eigenfunctions with nonzero eigenvalues of both diagonal generators are

$$|\beta, m\rangle = exp(i\beta arc\,tanh(\frac{z}{t}) + im\phi) = exp(i\beta arc\,tanh(v) + im\phi),$$

(A.3.b-6)

with eigenvalues β and m.

A more general eigenfunction is a product of this with a state of spin s, for any integral or half-integral value of s. Generator L_{zt}, here a differential operator on coordinates, does not act on the internal part of the statefunction.

A.3.c Eigenfunctions of rotation type

Another expression for the boost generator is obtained using the variable (which has no analog for the rotation generator because of the difference of signs),

$$\tau = z - t,$$

(A.3.c-1)

so

$$L_{zt} = i(z\frac{d}{dt} + t\frac{d}{dz}) = i(-z + t)\frac{d}{d\tau} = -i\tau\frac{d}{d\tau}.$$

(A.3.c-2)

As

$$L_{zt}exp(ip\tau) = p\tau exp(ip\tau),$$

(A.3.c-3)

exponentials are not eigenfunctions, which are

$$|n, m\rangle = \tau^n exp(im\phi) = (z - t)^n exp(im\phi).$$

(A.3.c-4)

So

$$L_{zt}|n, m\rangle = -in|n, m\rangle.$$

(A.3.c-5)

Since it is a label of a rotation group state, m must be integral. And as L_{zt} is a generator of the Lorentz group, n must also be integral. This would not necessarily be true if it belonged to an algebra of which it is the only generator.

A boost generator can be in the form of a momentum generator, translating the system in velocity space, which is what a boost is, or a rotation-type generator. The "rotation" is in z, t space, and actually moves the system from one path to another, one with different velocity. Note that angle ξ changes the orientation of a line from the origin to a point in this space, that is it changes the speed of the object from that at the first point to the speed at the second. For an actual rotation a transformation moves a system (or point) along a circle. But since the eigenvalue for a rotation in a space of definite metric is discrete, there is nothing equivalent to translation in momentum, or velocity, space, as there is for this non-definite metric space. Discreteness is necessary because a rotation of 2π is equivalent to the identity; there is no comparable requirement for hyperbolic functions.

Eigenfunctions $|\beta\rangle$ correspond to momentum eigenfunctions, the $|n, m\rangle$ ones to rotation eigenfunctions (spherical harmonics), which are sums of powers of the coordinates. There are two types of eigenfunctions as there are two forms of boost generators. The first type gives the functions of velocity that are not affected by the boost (although the value of the variable, the velocity, is), the second type gives the functions of coordinates. Rotations around z acting on a spherical harmonic leave the functional form invariant, but change the value of the angle, as happens for the second form of boost eigenfunctions.

A.3.d   Using recursion relations to find L_{zt} eigenfunctions

Another way of obtaining eigenfunctions of L_{zt} is using expressions for its matrix elements [Gel'fand, Minlos and Shapiro (1963), p. 188; Naimark (1964), p. 116] to find recursion relations. These are also eigenfunctions of L_z, $m = 0$, so that with the $m = 0$ basis vector of representation l of the rotation group (spherical harmonic Y_0^l) labeled χ^l,

$$L_{zt}\chi^l = lC_l\chi^{l-1} - (l+1)C_{l+1}\chi^{l+1}, \tag{A.3.d-1}$$

$$C_l = \frac{i}{l}\left(\frac{(l^2 - l_o^2)(l^2 - l_1^2)}{4l^2 - 1}\right)^{\frac{1}{2}}, \quad l = l_0, l_0 + 1, \ldots . \tag{A.3.d-2}$$

Thus

$$L_{zt}\chi^l = i\sqrt{\frac{(l^2 - l_o^2)(l^2 - l_1^2)}{4l^2 - 1}}\chi^{l-1} - i\sqrt{\frac{((l+1)^2 - l_o^2)((l+1)^2 - l_1^2)}{4(l+1)^2 - 1}}\chi^{l+1}. \tag{A.3.d-3}$$

Taking $l = l_o$ gives

$$\sqrt{\frac{(2l_o + 1)((l_o + 1)^2 - l_1^2)}{4(l_o + 1)^2 - 1}}\chi^{l_o+1} = i\Lambda\chi^{l_o}. \tag{A.3.d-4}$$

The required eigenfunction is

$$|\Lambda, m\rangle = \sum A_l\chi^l, \tag{A.3.d-5}$$

where the sum is over all l from l_o to infinity (for an infinite-dimensional Lorentz representation). Then

$$L_{zt}|\Lambda, m\rangle = \sum A_l L_{zt}\chi^l = i\sum A_l\sqrt{\frac{(l^2 - l_o^2)(l^2 - l_1^2)}{4l^2 - 1}}\chi^{l-1}$$

$$-i\sqrt{\frac{((l+1)^2 - l_o^2)((l+1)^2 - l_1^2)}{4(l+1)^2 - 1}}\chi^{l+1} = \Lambda|\Lambda, m) = \Lambda\sum A_l\chi^l, \quad \text{(A.3.d-6)}$$

and we have $L_{zt}\chi^l$ for each l, so

$$i\sum A_l\sqrt{\frac{(l^2 - l_o^2)(l^2 - l_1^2)}{4l^2 - 1}}\chi^{l-1} - i\sqrt{\frac{((l+1)^2 - l_o^2)((l+1)^2 - l_1^2)}{4(l+1)^2 - 1}}\chi^{l+1}$$

$$= \Lambda\sum A_l\chi^l, \quad \text{(A.3.d-7)}$$

plus the orthogonality of the χ^l, supplies a set of recursion relations for the A's, solvable for the set of (allowed) Λ's, giving the eigenfunctions and eigenvalues.

Here boost eigenfunctions are sums of spherical harmonics, a third form. Their variables are not those of the other two forms, being angles, θ, ϕ, in 3-space. There can be diverse realizations, such as functions over distinct spaces, using different variables, for the same set of operators.

We just note this form, for one reason to indicate the additional choice possible with a noncompact group (but do not mean to imply that there might not also be unusual ways (sec. III.3.a, p. 120) of realizing compact-group states and eigenstates). This diversity may not always be noticed, but should be.

A.4 Finite-dimensional Lorentz-group representations

The finite-dimensional representations are the spinors and have dimension [Gel'fand, Minlos and Shapiro (1963), p. 237; Naimark (1964), p. 127]

$$d_{mn} = (m + 1)(n + 1), \quad n \leq m, \quad \text{(A.4-1)}$$

where [Gel'fand, Minlos and Shapiro (1963), p. 238; Naimark (1964), p. 136]

$$l_o = \frac{1}{2}(m - n), \quad l_1 = \frac{1}{2}(m + n), \quad \text{(A.4-2)}$$

are the minimum and maximum angular momentum states in the representation.

For

$$m = 1, \quad n = 0, \quad l_o = l_1 = \frac{1}{2}, \quad d_{mn} = 2. \quad \text{(A.4-3)}$$

This is the two-component spinor.

Are there other cases for which the minimum and maximum angular momenta are the same? This requires

$$\frac{1}{2}(m - n) = \frac{1}{2}(m + n), \quad \text{(A.4-4)}$$

so $n = 0$ and the dimension is

$$d_{m0} = (m + 1) = 2l_o + 1. \qquad (A.4\text{-}5)$$

In general a spinor representation contains several representations of su(2). The familiar spin-$\frac{1}{2}$ representation of the Lorentz group is special.

A.5   Matrix elements of boost generators from a realization

The well-known derivations of the matrix elements of boost generators are algebraic [Gel'fand, Minlos and Shapiro (1963), p. 188; Naimark (1964), p. 116]. Since we have differential realizations of these it is fruitful to consider how they can be used to find matrix elements, and the limitations of realizations. These place conditions on operators, like their being derivatives, so are actually special cases — expressions for them may not be as general as for abstract operators to whose group or algebra their algebra is isomorphic. Abstract operators may, say, depend on arbitrary constants, which might be determined (or restricted) by the form of a realization — for only some values of these constants is a given realization possible. Expressions for boost operators emphasize this.

Also boost operators are found here from their action on spherical harmonics, and from this we construct Lorentz representations. But the nature of the expressions so obtained is independent of this use to which we put them, so their form is restricted by their being too specific, and thus not general enough. They are specific because their matrix elements are found from a derivative realization acting on specific functions, spherical harmonics. Therefore they can be put to other uses besides being Lorentz operators, so must be general, that is not restricted by this group. They are both too specific and too general.

Algebraic derivations treat operators as generators of a group, so they are limited by that and their matrix elements can contain, say, arbitrary constants. In fact, they must contain these for matrix elements depend on the Lorentz representation and are different for different representations — the constants are given by the representation labels. But with a realization, such as considered here, matrix elements are not found using the algebra, so cannot depend on algebra labels. These are fully determined from their action allowing no arbitrariness in their expressions.

A.5.a Calculation of matrix elements from action on coordinates

We need just L_{zt}, so the functions on which it acts can be taken to be in these variables only. This boost generator is one component of vector, that is one basis state of the adjoint representation [Mirman (1995a), sec. XIV.1.d, p. 405] of rotation group SO(3), and knowing it the other states (other boost generators) are given by the group, thus do not have to be considered.

The vector addition model [Mirman (1995a), sec. XII.2.a, p. 342] gives that the product of a vector with a state of representation l can be decomposed into a sum of states of representations $l + 1, l, l - 1$, as is well-known. A boost of spherical harmonic l, although not merely a product but also an operator acting on the function, gives a function that is a sum over states of representations $l + 1, l, l - 1$, as can be seen from the matrix elements.

A.5.b Matrix elements of rotation generators

How do we find matrix elements of L_{xz}? A basis function depends on x, y, z, and since it is a basis vector we can expand it and take term

$$|p, q, r\rangle = \frac{x^p y^q z^r}{\rho^3}. \tag{A.5.b-1}$$

A small rotation in the xz plane is given by

$$x \Rightarrow x + \theta z, \quad z \Rightarrow z - \theta x, \tag{A.5.b-2}$$

so (suppressing ρ which is fixed)

$$|p, q, r\rangle \Rightarrow |p, q, r\rangle' = (x + \theta z)^p y^q (z - \theta x)^r. \tag{A.5.b-3}$$

Since θ is small

$$|p, q, r\rangle' = |p, q, r\rangle + \theta y^q (p z^{r+1} x^{p-1} - r x^{p+1} z^{r-1})$$
$$= |p, q, r\rangle + \theta\{p|p - 1, q, r + 1\rangle - r|p + 1, q, r - 1\rangle\}, \tag{A.5.b-4}$$

$$|p, q, r\rangle' = exp(i\theta L_{xz})|p, q, r\rangle \Rightarrow |p, q, r\rangle + i\theta L_{xz}|p, q, r\rangle. \tag{A.5.b-5}$$

Thus

$$iL_{xz}|p, q, r\rangle = p|p - 1, q, r\rangle - r|p, q, r - 1\rangle. \tag{A.5.b-6}$$

For eigenfunctions of $L_z = L_{xy}$, $p = q = 0$. Then the terms have representation label $l = r$, and state label $m = 0$.

A.5.c Boost matrix elements

Similar considerations apply to L_{zt} which acts on $Y_0^l(w = \frac{z}{r})$. Coordinates x and y can be ignored and we use

$$r^2 = z^2 + \rho^2, \tag{A.5.c-1}$$

$$\rho^2 = x^2 + y^2, \tag{A.5.c-2}$$

with ρ a constant (under L_{zt}). Now for a small boost β,

$$z \Rightarrow z' = z + \beta t, \quad t \Rightarrow t + \beta z. \tag{A.5.c-3}$$

With βt, thus $\frac{\beta t}{r}$, small,

$$Y_0^l(w = \frac{z}{r} = cos(\theta)) \Rightarrow Y_0^l(w' = \frac{z'}{r'}), \tag{A.5.c-4}$$

$$r' = \sqrt{z^2 + \rho^2 + 2\beta tz} \Rightarrow r + \frac{\beta tz}{r}, \tag{A.5.c-5}$$

$$w' = \frac{z'}{r'} = \frac{z + \beta t}{\sqrt{z^2 + \rho^2 + 2\beta tz}} \Rightarrow (z + \beta t)(\frac{1}{r} - \frac{\beta tz}{r^3})$$

$$\Rightarrow \frac{z}{r} + \frac{\beta t}{r} - \frac{\beta tz^2}{r^3} = \frac{z}{r} + \frac{\beta t}{r}(1 - \frac{z^2}{r^2})$$

$$= w + \frac{\beta t}{r}(1 - w^2) = cos(\theta) + \frac{\beta t}{r}sin^2(\theta) = cos(\eta), \tag{A.5.c-6}$$

defining $cos(\eta)$,

$$Y_0^l(w' = \frac{z'}{r'}) = Y_0^l((z + \beta t)(\frac{1}{r} - \frac{\beta tz}{r^3}))$$

$$= Y_0^l(w = \frac{z}{r}) + \frac{\beta t}{r}\frac{dY_0^l((z + \beta t)(\frac{1}{r} - \frac{\beta tz}{r^3}))}{d\frac{\beta t}{r}}, \tag{A.5.c-7}$$

in the limit $\frac{\beta t}{r} \Rightarrow 0$. Thus taking the limit,

$$iL_{zt}Y_0^l(w) = \frac{d\{Y_0^l((z + \beta t)(r^{-1} - \beta tzr^{-3}))\}}{d\frac{\beta t}{r}}. \tag{A.5.c-8}$$

Now

$$Y_0^l\{(z + \beta t)(r^{-1} - \beta tzr^{-3})\} = Y_0^l(w + \frac{\beta t}{r} - \frac{\beta tz^2}{r^3}) = Y_0^l(cos(\eta)), \tag{A.5.c-9}$$

so

$$\frac{d\eta}{d(\frac{\beta t}{r})} = \frac{dcos(\eta)}{d(\frac{\beta t}{r})}\frac{d\eta}{dcos(\eta)} = -\frac{1}{sin(\eta)}(1 - \frac{z^2}{r^2}). \tag{A.5.c-10}$$

We need therefore

$$\frac{dY_0^l(cos(\eta))}{d(\frac{\beta t}{r})} = \frac{dY_0^l(cos(\eta))}{d\eta}\frac{d\eta}{d(\frac{\beta t}{r})}$$

$$= -\frac{1}{sin(\eta)}\frac{dY_0^l(cos(\eta))}{d\eta}(1-w^2).$$ (A.5.c-11)

So in the limit $\frac{\beta t}{r} \Rightarrow 0$,

$$iL_{zt}Y_0^l(w) = -\frac{1}{sin(\theta)}\frac{dY_0^l(cos(\theta))}{d\theta}(1-cos^2\theta)$$

$$= -sin(\theta)\frac{dY_0^l(cos(\theta))}{d\theta}.$$ (A.5.c-12)

Derivatives of spherical harmonics are known [Varshalovich, Moskalev and Khersonskii (1988), p. 147],

$$sin(\theta)\frac{dY_0^l(\theta,\phi)}{d\theta} = \frac{l(l+1)}{\sqrt{2l+1}}(\frac{Y_0^{l+1}}{\sqrt{2l+3}} - \frac{Y_0^{l-1}}{\sqrt{2l-1}}).$$ (A.5.c-13)

This gives

$$iL_{zt}Y_0^l(\theta) = \frac{-l(l+1)}{\sqrt{2l+1}}(\frac{Y_0^{l+1}}{\sqrt{2l+3}} - \frac{Y_0^{l-1}}{\sqrt{2l-1}}).$$ (A.5.c-14)

If the object has spin s its statefunction is

$$|l,s,j\rangle = Y_0^l(\theta)\chi_0^s C(j,0,l,0,s,0),$$ (A.5.c-15)

with the C the Clebsch-Gordan coefficient; the product $Y_0^l(\theta)\chi_{s0}$ is a basis state of a reducible representation. In general the product of two spherical harmonics transforms as a rotation group basis state [Mirman (1995a), eq. XII.2.b.ii-2, p. 345],

$$|L,0\rangle = C(L,0,l,0,\lambda,0)Y_0^l(\theta,0)Y_0^\lambda(\theta',0),$$ (A.5.c-16)

where C is the Clebsch-Gordan coefficient; these form an orthogonal set. Here it is reduced by taking traces and symmetrizing. However since Y_0^l is a sum of powers of z, this cannot be done so the only representation that appears in the decomposition is that for which $j = l + s$. We have the Clebsch-Gordan coefficients [Rose (1995), p. 40]. Thus

$$iL_{zt}|l,s,j\rangle = \frac{l(l+1)}{\sqrt{2l+1}}\sum(\frac{C(j,0,l+1,0,s,0)}{\sqrt{2l+3}}Y_0^{l+1} -$$

$$\frac{C(j,0,l-1,0,s,0)}{\sqrt{2l-1}}Y_0^{l-1})\chi_0^s.$$ (A.5.c-17)

The Y_0^{l-1} term is zero for $l = 0$. The smallest value of total angular momentum j is then s. For the Lorentz group this generates, from any rotation state, a set of states with j values differing by 1 from $j = s$ upward. Now

$$Y_0^{l+1}\chi_0^s = |l+1, s, j'\rangle C(j', 0, l+1, 0, s, 0), \qquad \text{(A.5.c-18)}$$

$$Y_0^{l-1}\chi_0^s = |l-1, s, j'\rangle C(j'', 0, l-1, 0, s, 0). \qquad \text{(A.5.c-19)}$$

So

$$\begin{aligned}
iL_{zt}|l, s, j\rangle &= \frac{l(l+1)}{\sqrt{2l+1}}\Big\{\frac{C(j+1, 0, l+1, 0, s, 0)}{\sqrt{2l+3}}|l+1, s, j+1\rangle \\
&\quad -\frac{C(j-1, 0, l-1, 0, s, 0)}{\sqrt{2l-1}}|l-1, s, j-1\rangle\Big\} \\
&= \frac{l(l+1)}{\sqrt{2l+1}}\Big\{\frac{C(l+s+1, 0, l+1, 0, s, 0)}{\sqrt{2l+3}}|l+1, s, l+s+1\rangle \\
&\quad -\frac{C(l+s-1, 0, l-1, 0, s, 0)}{\sqrt{2l-1}}|l-1, s, l+s-1\rangle\Big\}. \qquad \text{(A.5.c-20)}
\end{aligned}$$

Also

$$L_{zt}F^l = ilC_l F^{l-1} - i(l+1)C_{l+1}F^{l+1}. \qquad \text{(A.5.c-21)}$$

The known matrix elements, for a slightly different definition of the basis vectors which are arbitrary up to multiplication by functions of l, are [Gel'fand, Minlos and Shapiro (1963), p. 188; Naimark (1964), p. 116]

$$C_l = \frac{i}{l}\sqrt{\frac{(l^2-l_0^2)(l^2-l_1^2)}{4l^2-1}}, \quad l = l_0, l_0+1, \dots . \qquad \text{(A.5.c-22)}$$

These give

$$\begin{aligned}
iL_{zt}F_0^l(\theta) &= \sqrt{\frac{(l^2-l_0^2)(l^2-l_1^2)}{4l^2-1}}F_0^{l-1} - \sqrt{\frac{((l+1)^2-l_0^2)((l+1)^2-l_1^2)}{4(l+1)^2-1}}F_0^{l+1} \\
&= \frac{1}{\sqrt{2l+1}}\Big(\sqrt{\frac{(l^2-l_0^2)(l^2-l_1^2)}{2l-1}}F_0^{l-1} - \sqrt{\frac{((l+1)^2-l_0^2)((l+1)^2-l_1^2)}{2l+3}}F_0^{l+1}\Big).
\end{aligned}$$

$$\text{(A.5.c-23)}$$

For $l_0 = l_1 = 0$ the action of the boost is

$$iL_{zt}F_0^l(\theta) = \frac{1}{\sqrt{2l+1}}\Big(\frac{l^2}{\sqrt{2l-1}}F_0^{l-1} - \frac{(l+1)^2}{\sqrt{2l+3}}F_0^{l+1}\Big). \qquad \text{(A.5.c-24)}$$

A.5.d Comparison with the algebraic construction

Algebraic constructions are more general. There are two constants whose choice determines the Lorentz representation. But this realization does not contain constants, it is completely determined by its explicit form — matrix elements of the Lorentz representation are given by specific expressions for boost operators. Since the operators act on polynomials in z and are realized in terms of derivatives, they give zero only when the power of z is zero — the reason for the smallest l in the representation, $l_o = 0$. Matrix elements are essentially Clebsch-Gordan coefficients of the rotation group. These are fully determined so there is no way of putting in a second constant. And $l_1 = 0$. Why 0? The value of this is fixed by the matrix elements of the boost generators between states of the same rotation representation — it is defined as being proportional to these matrix elements; only L_{zt} is needed, which is all we consider. But the form of the realization of a boost generator prevents it from connecting states of the same representation: $l_1 = 0$.

Essentially the reason is that L_{zt} is a sum of $z\frac{d}{dt}$ — multiplication by z increases the value of l by 1 — and $t\frac{d}{dz}$, which decreases it by 1. The same is true for x and y, which also transform under the $l = 1$ representation. For l to remain the same there would have to be a term that removed one space variable and replaced it by another, like $x\frac{d}{dz}$. But this cannot be for there must be one t, so exactly one space coordinate is present.

Also this realization of the boost has odd parity [Mirman (1995b), sec. 6.2.2, p. 110], thus cannot have nonzero matrix elements between states of the same representation, this determining one of the labels, which as defined is zero. But there is no parity operator for symbols whose only property is that they obey commutation relations. Using only such symbols is more general, allowing other values for this label.

Thus the constants determining the Lorentz representation are fixed by the realization of states and operators, and at these values.

A.5.e   Representations whose basis states are plane waves

An alternate type of Lorentz group basis vector is a plane wave, which is relevant because it is a Poincaré basis vector with momentum diagonal. We just note the reason it is a Lorentz basis vector and how it is related to more familiar ones. Plane waves

$$|p_\mu, \phi\rangle = exp(ip_\mu x_\mu + i\phi), \quad \mu = 1, \ldots, 4, \qquad \text{(A.5.e-1)}$$

are Poincaré group basis vectors. For the Lorentz group there are no distances, so we can choose (the scalar product) $p_\mu x_\mu = 1$, limiting transformations to those of the Lorentz group. These leave p fixed and rotate coordinates x_j of 3+1 space. Each vector p_μ gives a state of the representation.

There are an infinite number of basis states. If we consider Lorentz operators to act on p_μ then each set of components gives a basis state, and components vary continuously. Likewise for operators acting on x_μ (for the space dual to that of p_μ), each set of its continuously varying values labels a different state (a different function of p_μ). Each set of components gives a distinct basis vector — operators of a group acting on a basis vector give a set of basis vectors, defining the specific representation. Here, acting on one state, one plane wave, one set of components, they generate all components of the set, all states of the representation.

Representations are labeled by the magnitude of $p_\mu x_\mu$, and are all (essentially) equivalent — units are arbitrary so there is no distinction between different values, except that they are different.

There is one frame in which all components of p_μ are zero except $p_4 = 1$, the rest frame. Or we can consider that all components of x_μ are 0, except $x_4 = 1$ — the object is at the origin and the value of "time" is 1. Since this is the Lorentz, not the Poincaré, group, time cannot vary and units are arbitrary, so this can be chosen.

A plane wave is expandable as a sum of (an infinite number of) spherical harmonics as we see next. All these appear from $l = 0$ to $l = \infty$. The representation is determined since this basis vector, the plane wave, has been completely specified (phase ϕ just gives an overall multiplicative factor).

We refer to representation consisting of these plane waves as continuous representations.

Basis states of the discrete representations (sec. A.2, p. 224) and those of continuous ones can be expanded in terms of each other — plane waves and spherical harmonics can be so expanded.

A.6 Expansion of Poincaré basis states in terms of Lorentz states

Basis states used here are those with the momentum diagonal, so are (sums of terms) of the form $exp(ip_j r_j - ip_4 t)$. Taking the unit of momentum as 1, this can be written $exp(i\rho f(\theta, \phi) - i\tau)$, with f a function of angles, ρ the magnitude of r, and $\tau = p_4 t$ in these units, a constant for the Lorentz group which does not involve time. It just gives an overall multiplicative factor so can be ignored. We study these representations by writing this form of its basis states in terms of those of the Lorentz group whose representations we know.

A.6.a Expanding a plane wave as a sum of spherical harmonics

The expansion of a plane wave in terms of spherical harmonics [Abramowitz and Stegun (1972), p. 440; Jackson (1963), p. 566; Panofsky and Phillips (1955), p. 205] is

$$exp(ip_\mu x_\mu) = 4\pi \sum_{l=0}^{\infty} i^l j_l(pr) \sum_{m=-l}^{l} Y_m^{l*}(\theta', \phi') Y_m^l(\theta, \phi), \quad (A.6.a\text{-}1)$$

where θ', ϕ' are the angles of p, and θ, ϕ those of x. This gives

$$exp(ip_\mu x_\mu) = \sum_{l=0}^{\infty} i^l \sqrt{4\pi(2l+1)} j_l(pr) Y_0^l(y), \qquad (A.6.a\text{-}2)$$

with y the angle between p and x, and j_l is a spherical Bessel function [Jackson (1963), p. 539]. This, with $J_{l+\frac{1}{2}}$ a Bessel function, is

$$j_l(q) = \sqrt{\frac{\pi}{2q}} J_{l+\frac{1}{2}}(q). \qquad (A.6.a\text{-}3)$$

For small values of the argument

$$j_l(q) \Rightarrow \frac{q^l}{(2l+1)!!}. \qquad (A.6.a\text{-}4)$$

Thus in the limit $q \Rightarrow 0$

$$j_l(q) \Rightarrow 0, \quad l > 0, \quad j_0(q) \Rightarrow 1. \qquad (A.6.a\text{-}5)$$

Also

$$j_0(q) = \frac{sin(q)}{q} \Rightarrow 1, \quad q \text{ small.} \qquad (A.6.a\text{-}6)$$

$$j_1(q) = \frac{sin(q)}{q^2} - \frac{cos(q)}{q} \Rightarrow \frac{1}{q} - \frac{1}{q} = 0. \qquad (A.6.a\text{-}7)$$

Another form is

$$exp(iprcos(\theta)) = \sum_{l=0}^{\infty} i^l (2l+1) j_l(pr) P_l(cos(\theta)). \qquad (A.6.a\text{-}8)$$

Taking p along z,

$$exp(ip_\mu x_\mu) = exp(ip_z rcos(\theta)cos(\phi)) = 4\pi \sum_{l=0}^{\infty} i^l j_l(pr) \sum_{m=-l}^{l} Y_m^l(\theta, \phi).$$
$$(A.6.a\text{-}9)$$

Spherical Bessel functions obey orthogonality relations [Arfken (1970), p. 527]

$$\int_{-\infty}^{\infty} j_m(q) j_n(q) dq = 0, \quad m \neq n, \quad 0 \leq m, n, \qquad (A.6.a\text{-}10)$$

$$\int_{-\infty}^{\infty} j_m(q) j_m(q) dq = \frac{\pi}{2m+1}. \qquad (A.6.a\text{-}11)$$

A.6.b   Spherical harmonics as sums over plane waves

Conversely spherical harmonics can be expanded in terms of plane waves. Multiply eq. A.6.a-1, p. 240, by $sin(\theta')Y_{m'}^{l'}(\theta',\phi')$, integrate over θ',ϕ', and use the orthogonality of spherical harmonics, which holds because these are representation basis vectors [Jackson (1963), p. 65; Mirman (1995a), sec. VII.5.b, p. 195; sec. XI.4.c, p. 322]. Then

$$Y_m^l(\theta,\phi) = \frac{1}{4\pi i^l j_l(pr)} \int d\Omega' sin(\theta')Y_{m'}^{l'}(\theta',\phi')exp(ip_\mu x_\mu)$$

$$= \frac{1}{4\pi i^l j_l(pr)} \int d\Omega' sin(\theta')Y_{m'}^{l'}(\theta',\phi')$$
$$\times exp(ipr\{cos(\theta)cos(\phi)cos(\theta')cos(\phi')\}).$$
$$(A.6.b-1)$$

Also writing $j_n(q)$, $q = p_\mu x_\mu$, we have, using eq. A.6.a-11, p. 241,

$$\int_{-\infty}^{\infty} j_n(q)exp(iq)dq$$

$$= 4\pi \sum_{l=0}^{\infty} i^l \int_{-\infty}^{\infty} j_n(q)j_l(q)dq \sum_{m=-l}^{l} Y_m^{l*}(\theta',\phi')Y_m^l(\theta,\phi)$$

$$= \frac{4\pi^2 i^n}{2n+1} \sum_{m=-n}^{n} Y_m^{n*}(\theta',\phi')Y_m^n(\theta,\phi).   (A.6.b-2)$$

If we set $\theta',\phi' = 0$,

$$\sum_{m=-n}^{n} Y_m^n(\theta,\phi) = \frac{(2n+1)}{4\pi^2 i^n} \int_{-\infty}^{\infty} j_n(q)exp(iq)dq.   (A.6.b-3)$$

From

$$exp(iq) = \sum_{l=0}^{\infty} i^l j_l(q)Y_0^l(y),   (A.6.b-4)$$

we get

$$Y_0^l(y) = \frac{(2n+1)i^n}{\pi} \int_{-\infty}^{\infty} j_n(q)exp(iq)dq.   (A.6.b-5)$$

A.7   The SO(2,1) subgroup of the Lorentz group

Matrix elements of SO(2,1) can be obtained from those of the Lorentz group, of which it is a subgroup [Naimark (1964), p. 104], but since it is a subgroup the representation spaces are larger than those defined by the operators of SO(2,1). This has implications, perhaps quite fundamental ones, that remain to be explored. All that we can do here is point this out in the hope of stimulating further research.

The subgroup consists of H_3 (taken diagonal), and boost genera-tors F_+ and F_-, the noncompact Lorentz generators changing ("total angular momentum") k (often denoted as S, L or J) of the rotation sub-representation of the Lorentz-group representation. States are given by Lorentz-group-representation labels k_o and c, which determine the SO(2,1) representation (so matrix elements), and by state label v, the eigenvalue of H_3, and k. Although SO(2,1) has only one invariant there are two representation labels, and also two state labels, k and v. Thus there are labels that are not eigenvalues of algebra and enveloping alge-bra operators of the subgroup. As these representations are realized on a larger space than that defined by SO(2,1), the SO(2,1) canonical defini-tion space, which is the 2+1-dimensional nondefinite-metric real space whose transformations define the group elements (sec. III.5.b, p. 162), there are more representations and states, so the subgroup is not able to label them completely.

SO(2,1) representations subduced from those of the Lorentz group irreducible representation [Mirman (1999), sec. VI.2, p. 281] are them-selves irreducible. On a set of states all with the same k (those belong-ing to a single rotation subrepresentation of the Lorentz representa-tion) the F operators (boost generators) have the same action as the H's (rotation generators) do (however not the same matrix elements, which for these are found using the Wigner-Eckart theorem [Mirman (1995a), sec. XII.3.b, p. 350]) — but in addition these mix rotation rep-resentations. Thus all states of the Lorentz representation are reached by these subalgebra operators. If it were possible to take linear combi-nations of states invariant under the subalgebra then these same linear combinations would be invariant under the Lorentz group (since matrix elements of the other boost generators differ only in their dependence on the rotation labels, and rotation operators connect all states of each rotation representation), so its representation would not be irreducible, as is assumed. Undoubtedly a rigorous proof can be given, which we cannot consider here.

Note the difference between this and the decomposition with respect to the rotation group for which the Lorentz representation consists of a set of rotation representations and the transformations of the rotation group take a state into other states, but of the same rotation representa-tion. Lorentz transformations from outside the rotation group, boosts, take a state of one representation into states of other rotation represen-tations. But transformations of this SO(2,1) subgroup take any state of the Lorentz representation into any other state — these subgroup trans-formations (iterated) connect all states of the Lorentz representation, even though they are transformations of only a subgroup.

A.8   Noncompact groups and subgroups

The Lorentz and SO(2,1) groups are noncompact. Also the subalgebra is realized on the representation space of the algebra, which is larger than the space defined by the action of the subalgebra elements alone. Thus it is clear that we cannot conclude that representations and states of a group are always labeled by eigenvalues of algebra and enveloping algebra operators of that group alone. Representations can be realized on larger spaces, with more states, and these are too many for the subalgebra to label. They are determined rather by values of matrix elements of the operators between them. We leave open the questions whether there are simple ways of giving these extra labels, whether there are operators related to them?

Since SO(2,1) can also be regarded as a subgroup of an infinite number of larger groups undoubtedly its representations can also be realized on much larger spaces, requiring more labels which it is unable to supply. There might well be in addition other realizations and some of these, and their states, might not have any labels (sec. III.3.a, p. 120) that are eigenvalues [Mirman (1995c), sec. 2.5.2, p. 26], or eigenvalues may not distinguish different representations.

This emphasizes again that we must be careful in generalizing to others what we have learned from (linear realizations of) compact, semisimple groups.

A.9   Nonlinear realizations and nonlinear differential equations

For groups like the conformal group, realizations of some elements are nonlinear (sec. III.5.b, p. 162). This might be useful, say, in finding solutions of nonlinear differential equations. Generators of a group can also be realized as differential operators over spaces larger than those defined by the group itself, such as representation spaces of groups of which it is a subgroup, as we see. This leads to the interesting question of the form of nonlinear realizations over such larger spaces? Might these also be useful, perhaps in finding solutions of nonlinear differential equations or special functions? Might these give solutions that cannot be found from smaller spaces, perhaps of equations that cannot be solved using smaller spaces? There are many such questions and possibilities raised by combinations like those of nonlinear realizations and larger representation spaces. Perhaps there are other options suggested by these groups and realizations that would be useful, including those of various combinations.

Such possibilities are also interesting because nonlinear forms of operators result from defining a group over a space smaller than its canonical definition space, while these representations are found by having these operators acting on larger spaces than generated by the operators themselves. The four series of classical groups are defined by their action on complex, real, and symplectic spaces, giving the unitary, orthogonal, and symplectic groups. These complex, real, and symplectic spaces are their canonical definition spaces for they are the spaces used to define the groups and determine their structure. Mixtures of spaces of both fewer and of more dimensions could prove useful, and we have seen many hints of this. The conformal group (of 3+1 space) is defined over a space whose dimension, of four real coordinates, is less than that of the canonical definition spaces, with four complex or six real coordinates. Action of conformal generators on basis vectors of (the larger) representation spaces obtained using operators of such homogeneous groups over these canonical definition spaces remains to be investigated, and likely should be.

In particular, it may not be too difficult to obtain representations of simple groups, and they can be expected to be functions of simple functions, like *sine*'s and *sinh*'s. Relating representations of nonlinearly-realized groups, like conformal groups, to these might give representations of the latter with perhaps not too much difficulty. If these representations can be related to solutions of nonlinear differential equations, the approach would have the possibility of obtaining such solutions, if not in closed form, at least as effective expansions. Certainly investigations of these is reasonable.

A.10   Increasing complexity and richness

Of course the conformal group illustrates many such interesting and suggestive possibilities. For larger simple groups the number of spaces of smaller dimension on which they can be realized increases, giving perhaps more interesting (higher-order?) nonlinearities, and increased structure. Also the number and types of decompositions of representations into representations of subgroups grow rapidly, suggesting that there might be much richness, and perhaps many applications, obtained with these kinds of analyses.

They imply again that usual approaches to and assumptions about group representations are restrictive and that realizations and representations, linear and nonlinear, differential, perhaps others, can be far more abundant than is perhaps usually believed (sec. III.5.d, p. 164).

Vast fields for exploration may exist, too vast to be studied here, but whose valuable possibilities, of many different types, make such explorations intriguing.

Appendix B

What information do elementary particle masses give?

B.1 The importance of spectroscopy and the elementary particle masses

Spectroscopy has historically been of major importance as a tool for understanding physical systems. And so it is for elementary particles; much can be learned about them from their masses. There is a simple formula to which they (many?) conform. This is studied, as are isospin splittings, and both the agreement and the discrepancies provide much information and many clues. Unfortunately the significance of these is far from clear.

It has been well-known for many years [Mirman (1963)] that the masses of the elementary particles conform to a simple rule. However many new particles have been found since that was pointed out (if indeed that was the first time). Thus reviewing the extent to which these new values are in accord with the rule, and the information, hints, clues, that can be obtained from the masses, including the new values, is likely to be rewarding. There are in fact some quite suggestive clues. Regrettably it is not clear what they suggest. That is a major reason why it is useful to return to the mass-level formula. Perhaps there are readers who will be able to extract more information, and, hopefully, put the information to good use.

This topic has a long history. The first person to notice that the masses are given by $\frac{n}{\alpha}$ apparently was Nambu [Nambu (1952)], a remarkable observation considering how few particles were known at the time. Others who have discussed this or related concepts are Levitt

246

[Levitt (1958)], who also gives other references; Fröhlich [Fröhlich (1958, 1960)]; Umezawa [Umezawa (1964)]; Pease [Pease (1970)]; Becker [Becker (1974)] and MacGregor [MacGregor (1978)]. There are likely to have been others.

The masses of (charged) particles, in terms of the electron mass, with few exceptions, lie close to the levels given by

$$m = n(\frac{1}{\alpha} + a), \tag{B.1-1}$$

where n is an integer or half-integer,

$$a = 0, \pm 1, \tag{B.1-2}$$

and α the fine-structure constant. The involvement of the fine-structure constant is one of the great mysteries of particle physics.

These are the allowed levels, but most are unoccupied, and it is not clear what determines the occupied ones. If a rule could be found, it should be enlightening and significant.

This does not hold for the electron, but the masses are expressed in terms of its mass, nor the neutrinos (at least the electron neutrino whose mass is strongly bounded) if they are not massless (and it is difficult to believe that they can have zero mass [Mirman (1995c), sec. 4.4.4, p. 70; sec. 8.1.6, p. 144]). However none of the masses fall exactly on these levels, and perhaps for neutrinos $n = 0$, with their masses the (necessary) deviations.

The formula gives the allowed mass levels for all particles, not, as some might expect, the mass-squared for mesons.

There is an interesting, and strange, pattern of deviations, at least for the lowest octets and decuplets. The masses of charged particles (or at least of one of an isospin multiplet) fit quite well, those of the neutral ones, and also the charged hadrons with isospin zero, are in lesser agreement (although in general better than chance). Within an isospin multiplet it is a charged member that agrees with the formula. This is the exact opposite of what might be expected from perturbation theory. However there are other reasons for being dubious about drawing conclusions from perturbation theory [Mirman (1995b), sec. 1.6, p. 20; (1995c), sec. 11.2.2, p. 187; (2001)].

B.2   The table comparing experiment and the formula

In the table (tbl. B.9, p. 256) are listed the accepted particles with all relevant experimental values [Particle Data Group, (2000)] with experimental error in mass of 15 Mev or less. It is meaningless to consider particles whose masses are not well-known. This cutoff is arbitrary but one has to be chosen. Of course, values with large uncertainties should not be taken seriously (unless there is reason to believe that the errors are too conservative). For a few cases the existence of the resonance is uncertain — it is stated that it needs confirmation — so these are not included.

The particle names are those given by the Particle Data Group (2000), p. 84, except that there many include the masses which are not listed here since these are in the following column. That there is more than one particle with the same symbol should not cause confusion. Listed are the particle names, and charges (if given and not zero), experimental mass values, the closest value given by the formula in Mev, than in terms of the electron mass, which is clearly the proper unit, the difference in Mev, and the n and a values. (It would likely lead to greater understanding if masses were always given in terms of the electron mass.) Unless the charge is stated, the value is zero or for a mixture of charge states, of course making comparisons difficult.

The levels are calculated using [Particle Data Group, (2000), p. 73]

$$\frac{1}{\alpha} = 137.03599976 \ (= 137.03600) \tag{B.2-1}$$

taking it to the same number of decimal places as the muon mass (the most precise one). The conversion factor between electron mass and Mev is [Particle Data Group, (2000), p. 313]

$$m_e = 0.510998902 \ (= 0.51100) \ \text{Mev}. \tag{B.2-2}$$

In a few cases there are two levels close to the experimental mass. It is possible that for some this might not be accidental. For the muon, because of the smallness of n, there are two a values that give close agreement, although one is substantially better. Both charged particles of the Σ (1190) triplet lie close to levels given by the formula, but to adjacent ones. This is suggested again for the other Σ triplet, for which agreement is poorer. Possibly this is significant.

Should experimental error not permit a clear choice of the closest level, both are listed. Some levels may have more than one occupant, but this cannot be verified because of experimental error so occasionally two levels are given since the particles would be on adjacent ones if it should turn out that there can be only one particle (one isomultiplet) per level.

Because the mass-level formula is expected to be in very close agreement with only a subset of particles (a well-defined one), the differences between the levels and the experimental values for these particles are underlined. Most, but not all, are in excellent agreement. These are the values that especially emphasize the relevance of the formula to particle physics.

B.3   Disagreements and implications

The agreements of the Ω, two of the Λ_c^+'s, and the D_S's, are poor. Perhaps it is relevant that though these are charged, they have isospin 0. In fact, almost all particles with isospin 0 (total or third component) are in poorer agreement than charged particles with nonzero isospin, a pattern that might be meaningful. Perhaps some of the deviations are due to mixing, and it could be easier for such particles.

For a few other cases, especially the τ, $\Sigma^-(1387)$, and $\Xi^-(1535)$, agreement is poor, with the differences greater than about 1 Mev (although most are better than chance; in particular there are very few particles with large deviations). For several cases the differences are less than two standard deviations from 1 Mev; these might be due to experimental error. The worst case seems to be the $K^{*\pm}(891)$; also for this the neutral particle might lie closer to a level than the charged one. A possibility is that the formula gives the first two terms of an expansion (which seems reasonable), and that for this particle the other terms are "accidently" large. Also there might be some peculiarity of the particle that experiment has not picked up. Perhaps there is mixing, but that would raise the question why there is for this but not (as much) for others. But it could be that the deviation from the mass-level formula is significant and telling us something important about either the particle or particles in general. In other cases the disagreement could be due to experimental error, so it is only the $K^{*\pm}(891)$ for which there is definite disagreement. The deviations for all members of the 1^{--} nonet (ρ, K^*, ϕ, ω) to which the $K^{*\pm}(891)$ belongs [Particle Data Group (2000), p. 118] are relatively large. It is perhaps noteworthy that there are other 1^{--} nonets. There are also other 0^{-+} nonets besides the lowest one (π, K, η, η'), but these lie further away.

The agreement for several $b\bar{b}$ particles is unusually poor. However several are candidates for mixing, which is perhaps not irrelevant.

The best agreements, except for (the far too many) particles whose experimental errors are large enough to preclude definite statements, are for the charged, nonzero-isospin, particles (but including the leptons) with the longest lifetimes (those for which the lifetimes are listed [Particle Data Group (2000)]; these starred in the table), using the best value within an isospin multiplet. They all have discrepancies of about 1 Mev or less, usually much less. The $B(5279)$ is an exception. There are enough examples to suggest a pattern, but not enough to be sure (especially about whether it is significant). This could also be an experimental artifact. For particles with measurable widths, it may not be possible to determine the relevant exact value of the mass, and so the other particles might generally be in better agreement than it seems. Otherwise there is a hint that being a long-lived charged member of an isospin multiplet anchors a particle to a level, but others float somewhat; also both zero isospin and zero charge cause the mass to change. More, and more precise, mass values would be useful.

For the Δ, both the masses and the pole positions are given [Particle Data Group (2000)]. Clearly the masses, not the pole positions, are in agreement.

It is interesting that deviations from the mass-level formula are often less, even much less (1 or 2 Mev), than the experimental errors. Could some be too conservative? Indeed, while the number of fairly large disagreements for particles that we expect to have small ones is few, the number of small (even very small) disagreements for others is surprisingly large. It is possible that in some cases this is not an accident, and with further data other patterns will emerge, and that those seen for the smallest mass particles might not hold in general.

Why is the agreement of the $\chi_{c1}(1P)(3510)$ so good?

The deviations thus hint at important clues and regularities, but the data are insufficient to extract them.

B.4   The formula must be relevant

How can we be sure that the formula is relevant to elementary particle physics? The agreement is extremely good (in cases for which a clear-cut statement can be made) almost always less than about an Mev, for one (the kaon) less that 0.1 Mev. Indeed the μ and π alone demonstrate the relevance of the $\frac{n}{\alpha}$ term. The cases with poor agreement (especially ones with small enough experimental error to allow a definite statement) are few. Importantly, there are no arbitrary parameters and no way of adjusting the levels. Also the pattern that the neutral particles (and apparently zero-isospin charged ones) deviate and the charged ones of a multiplet (plus the leptons) conform is a strong indication that the concurrence is not random (something that is unbelievable, in any case, from the close agreement). And most particles have been discovered since the formula was initially stated. These new particles are also in excellent agreement, indeed generally as good as the older ones. The formula made many predictions (incomplete because there is no known rationale for which levels are filled), and these have (almost always) been experimentally verified.

B.5   What is it telling us?

Thus the relevance seems certain. What can we learn?

It is interesting that there is no distinction between particles of different quantum numbers (except for the neutral shift, and probably zero isospin; it seems that particles with $I_z = 0$ are shifted, no matter what their value of I is). Almost all particles (including these shifted ones) lie near the levels, fermions and bosons, of all spins and isospins, leptons and hadrons, having small or large mass, with and without strangeness, charm and so on. The n and a values seem independent of the properties of the particles. Whether n is integral or half-integral does not appear related to any particle characteristic. There seems no way of telling from the level on which a particle lies, or its deviation, anything about it (except an indication of charge — is this related to α being in the formula?). Strongly interacting particles fit as well as particles with no strong interactions. The levels seem to be some universal property of nature on which the particles must lie (with small deviations).

It seems that a values of 1 and -1 occur more often than 0, for reasons that are not clear, and that may or may not be significant.

If deviations are partially due to mixing, if the "correct" particles lie close but some experimentally measured ones do not, then the formula gives a way of studying this, likely important, aspect. More and better mass values could be used to explore the question, and answers might be quite valuable.

B.6   There are two formulas

There are two formulas for the masses, this and that of Gell-Mann–Okubo [Cornwell (1984), p. 755; Schensted, p. 310]. Are they related? For the baryon octet, taking the n values of any three members, the Gell-Mann–Okubo formula does correctly give the n value for the fourth. This is not true for the a values, nor for the decuplet, nor for the mesons. Thus the agreement for the octet may be accidental. However if this should hold for others (for which too few masses are known to say anything) that would be notably informative. Otherwise it is strange that there are two formulas, in excellent agreement with experiment, that (appear to) have nothing to do with one another. This, for which nothing can be said without more experimental results, provides a major clue.

B.6.a   The expression of Gell-Mann and Okubo

The Gell-Mann–Okubo mass formula,

$$m = m_o + aY + b[I(I + 1) - \frac{Y^2}{4}], \qquad (\text{B.6.a-1})$$

where Y is the hypercharge and I the isospin, has the unattractive property that Y changes sign for antiparticles although the mass remains the same (by the TCP theorem, a trivial consequence of rotational invariance in an even dimensional space [Mirman (1995c), sec. 4.2.6, p. 60]).

A perhaps better alternative, giving the same masses (for the known particles), is

$$m = m_o + aBY + b[I(I+1) - \frac{Y^2}{4}], \tag{B.6.a-2}$$

with B the baryon number. If this is actually more appropriate it may indicate something theoretically.

B.6.b   Implications of the two formulas

Of course α must have a value that allows both formulas to be satisfactory simultaneously. However it is difficult to see where this leads as neither holds exactly, and we have no information about expressions for the deviations. But if the masses of more multiplets were known, they might provide insight.

This suggests part of the reason for the occupied levels and also for the values of the constants in the Gell-Mann–Okubo formula. It is only with some values that both can hold (to a reasonable approximation) simultaneously. And it hints at one reason for which particles (can) exist.

The Gell-Mann–Okubo formula gives the mass as a function of the state labels [Mirman (1995a), sec. I.6, p. 25]. Perhaps n and a are also state labels. But of what states, of what groups?

Might the other coupling constants appear in the formula? The value of the strong coupling constant [Particle Data Group (2000)] is about 0.1 (but it is really conventional), and the deviations are of that order or several times it. Thus these could depend on this value. Since the reciprocal of the square of the electromagnetic coupling constant appears we should consider these for the strong coupling constant. Its square, about 0.01 would be too small to be discernible, while its reciprocal, about 10, is of the order of some deviations. The square of this, about 100, is roughly $\frac{1}{\alpha}$, thus should have a major effect of which there is no indication. The gravitational and Fermi (weak interaction) constants are (in terms of the electron mass) of order 10^{-44}, and 10^{-11}, respectively. Contributions from them could not be determined. However the latter is really determined by the mass of the intermediate boson, the W mass. The correct pure number is (sec. IV.3.d.ii, p. 207)

$$g_w = 0.2. \tag{B.6.b-1}$$

The appearance of this cannot be ruled out.

If the reciprocal of the gravitational constant were used it would gives particles with such large masses that if they could be produced they would have been noticed. Their squares would be worse. So for it we can say nothing (except that their inverses and squares are not likely to be relevant).

The fine structure constant is definitely involved, the strong coupling constant, or its reciprocal, and the Fermi constant, are possibilities. Thus whether the fine structure constant is in some sense more fundamental than the others, or whether it just has the right value to be noticed, is unclear.

For mesons, terms linear in Y cannot occur for SU(3) multiplets that have both particles and antiparticles. For the formula to be checked experimentally, since there seems to be mixing of the neutral members of a nonet, the mixing angle must be known independently. Thus no clear statements can be made for mesons. Likewise terms linear in I_3 cannot occur for multiplets containing both particles and antiparticles. There is no evidence that it does (actually there is no evidence of any simple relationships for isospin splitting).

B.7   Masses and isospin

Another aspect of the masses, not (apparently) related to this formula, is the mass splittings within an isospin multiplet, which have several interesting, but highly unclear, properties. These are given in the second table (tbl. B.10, p. 268). It might be thought that the splittings are proportional to the eigenvalue of I_3, the diagonal isospin generator. For many mesons the positive and negative particles are TCP pairs, so must have the same mass (and it is interesting that it is possible for an isospin multiplet to contain such pairs, rather then having them form two multiplets so related, and that this does occur). So we must look to the baryons. Clearly, and perhaps surprisingly, the masses are not even roughly proportional to I_3. Might they at least be monotonic? There are two counterexamples, the Σ_c and perhaps the Δ. However in both cases the experimental errors are so large as to prevent a definitive statement. If these errors could be reduced, it would be interesting, and likely quite instructive.

Also the SU(3) mass splitting is given by the neutral operators (Y, $I(I+1) - \frac{1}{4}Y^2$), but the splitting within an isospin multiplet seems to have nothing to do with the isospin neutral operator (I_3).

It might be relevant that the splittings are all roughly about the same size, 3 – 5 Mev; the deviations of the neutral particles from the mass-level formula are also, generally, of the same magnitude (something to note). There is only one case, the Ξ, in which the splitting is definitely much larger, and a couple for which it might be. There are a few baryons in which it is clearly much smaller (really only the nucleon and the two Σ's). It might also be expected that the ratio of the splitting to the mass would be about constant. This obviously is not true. In fact the lightest particle, the pion, has, except for two cases, about the largest definite splitting. And the smallest mass difference within a multiplet might be for the $\Sigma(2453)$, about the heaviest particle for which there is data (although because of the large errors it is not clear which actually is the smallest). Thus with a few exceptions, the shift of mass because a particle is neutral (or apparently of zero isospin) is about 3 ± 1 Mev (6 ± 2 electron masses). Perhaps interestingly some of the lighter mesons seem to have larger shifts.

If the deviations of the neutral particles (and perhaps those of isospin zero) are due to mixing (or are only possible, or larger, for these) then that they are all roughly equal seems strange (at least unexpected).

B.8   Conclusion and implications

This formula is in accord with the strongly-expressed philosophy [Mirman (1995a,b,c;2001)] that the basic laws of physics, and their reasons, are simple, even trivial, surprisingly so. Certainly the formula is, even if the deviations are (apparently) not.

The particle masses provide us with much information, undoubtedly of great significance. But what this is all saying is quite unclear. One problem is that so few of the masses are known precisely enough. Indeed the imprecision of many is quite surprising. More experimental work, more masses, greater precision, should prove rewarding. Yet we do have much of value. Possibly readers will gain understanding from it. These are some experimental facts, along with many others, that a theory of elementary particles must explain. They provide clues, and puzzles.

B.9 Table for masses

Table B.9: Comparison of experimental and predicted masses

Particle	$I^G(J^{PC})$	Experimental Mass = M_x	Mass Level = M_l (Mev)	Mass Level (electron mass)	ΔM (Mev) = $M_x - M_l$	n	a
LEPTONS							
e	$\frac{1}{2}$	0.51099890 ± 0.000000021	1				
μ	$\frac{1}{2}^*$	105.658357 ± 0.000005	105.8046	207.054	-0.1462	1.5	1
			105.0381		0.62027	1.5	0
τ	$\frac{1}{2}^*$	1777.0 ± 0.3	1772.6	3468.9	4.4	25.5	-1
MESONS							
S = C = B = 0							
$\pi^{\pm}$	$1^-(0^-)^*$	139.57018 ± 0.00035	140.0508	274.072	-0.4806	2	0
π^o	$1^-(0^{-+})^*$	134.9766 ± 0.0006			-5.0742		
η	$0^+(0^{-+})$	547.30 ± 0.12	556.12	1088.29	-8.82	8	-1

APPENDIX B. WHAT INFORMATION DO ELEMENTARY PARTICLE MASSES GIVE?

Particle	$I^G(J^{PC})$	Experimental Mass = M_x	Mass Level = M_l (Mev)	Mass Level (electron mass)	ΔM (Mev) = M_x - M_l	n	a
$\rho^\pm$	$1^+(1^{--})$	766.5 ± 1.1	770.3	1507.4	<u>-3.8</u>	11	0
ρ^0	$1^+(1^{--})$	769 ± 1	770				
ω	$0^-(1^{--})$	782.57 ± 0.12	775.90	1518.40	6.7	11	1
η'	$0^+(0^{-+})$	957.78 ± 0.14	952.24	1863.49	5.5	13.5	1
f_o	$0^+(0^{++})$	980 ± 10	980	1919	0	14	0
a_o	$1^-(0^{++})$	984.8 ± 1.4	987.5	1932.5	-2.7	14	1
ϕ	$0^-(1^{--})$	1019.417 ± 0.014	1022.778	2001.522	-3.36	14.5	1
b_1	$1^+(1^{+-})$	1229.5 ± 3.2	1225.4	2398.1	4.1	17.5	0
f_2	$0^+(2^{++})$	1275.4 ± 1.2	1269.7	2484.6	5.7	18	1
f_1	$0^+(1^{++})$	1281.9 ± 0.6	1286.0	2516.7	-4.1	18.5	-1
η	$0^+(0^{-+})$	1297.0 ± 2.8	1295.5	2535.2	1.5	18.5	0
a_2	$1^-(2^{++})$	1318.0 ± 0.6	1320.8	2584.7	-2.8	19	-1

Particle $I^G(J^{PC})$	Experimental Mass = M_X	Mass Level = M_l (Mev)	Mass Level (electron mass)	ΔM (Mev) = M_X - M_l	n	a
f_1 $0^+(1^{++})$	1426.3 ± 1.1	1425.0	2788.7	1.3	20.5	-1
f_0 $0^+(0^{++})$	1500 ± 10	1505.5	2946	-6	21.5	0
		1494.6	2925	-6	21.5	-1
f_2' $0^+(2^{++})$	1525 ± 5	1529	2993	-4	22	-1
f_2 $0^+(2^{++})$	1638 ± 6	1634	3197	4	23.5	-1
η $0^+(2^{-+})$	1632 ± 14	1634	3197	-2	23.5	-1
ω_3 $0^-(3^{--})$	1667 ± 4	1668	3265	-1	24	-1
ρ_3 $1^+(3^{--})$	1691 ± 5	1693	3313	-2	24	1
f_0 $0^+(0^{++})$	1715 ± 7	1716	3357	-1	24.5	0
π $1^-(0^{-+})$	1801 ± 13	1799	3520	2	25.5	1
ϕ_3 $0^-(3^{--})$	1854 ± 7	1856	3631	-2	26.5	0
a_4 $1^-(4^{++})$	2014 ± 15	2016	3945	-2	29	-1
f_4 $0^+(4^{++})$	2034 ± 11	2031	3974	3	29	0

APPENDIX B. WHAT INFORMATION DO ELEMENTARY PARTICLE MASSES GIVE?

Particle $I^G(J^{PC})$	Experimental Mass = M_x	Mass Level = M_l (Mev)	Mass Level (electron mass)	ΔM (Mev) = $M_x - M_l$	n	a
S = ± 1, C = B = 0						
K± $\frac{1}{2}(0^-)$*	493.677 ± 0.016	493.755	966.252	-0.078	7	1
K⁰ $\frac{1}{2}(0^-)$*	497.672 ± 0.031			3.917		
K*± $\frac{1}{2}(1^-)$	891.66 ± 0.26	881.71	1725.45	9.95	12.5	1
K*⁰ $\frac{1}{2}(1^-)$	896.10 ± 0.27	903.69	1768.47	-7.59	13	-1
K₁ $\frac{1}{2}(1^+)$	1273 ± 7	1270	2485	3	18	1
K₁ $\frac{1}{2}(1^+)$	1402 ± 7	1401	2741	1	20	0
K₀* $\frac{1}{2}(0^+)$	1412 ± 6	1411	2761	1	20	1
K* $\frac{1}{2}(1^-)$	1414 ± 15	1411	2761	3	20	1
K₂*± $\frac{1}{2}(2^+)$	1425.6 ± 1.5	1425.0	2788.7	0.6	20.5	-1
K₂*⁰ $\frac{1}{2}(2^+)$	1432.4 ± 1.3			7.4		
K₂ $\frac{1}{2}(2^-)$	1773 ± 8	1773	3469	0	25.5	-1

Particle $I^G(J^{PC})$	Experimental Mass = M_x	Mass Level = M_l (Mev)	Mass Level (electron mass)	ΔM (Mev) = $M_x - M_l$	n	a
K_3^* $\frac{1}{2}(3^-)$	1776 ± 7	1773	3469	3	25.5	-1
K_2 $\frac{1}{2}(2^-)$	1816 ± 13	1821	3563	-5	26	0
K_4^* $\frac{1}{2}(4^+)$	2045 ± 9	2046	4003	-1	29	1
$C = \pm 1$						
$D^\pm$ $\frac{1}{2}(0^-)^*$	1869.3 ± 0.5	1869.2	3658.0	0.1	26.5	1
D^0 $\frac{1}{2}(0^-)^*$	1864.5 ± 0.5			-4.7		
$D^{*\pm}$ $\frac{1}{2}(1^-)$	2010.0 ± 0.5	2010.3	3934.0	-0.3	28.5	1
D^{*0} $\frac{1}{2}(1^-)$	2006.7 ± 0.5			-3.6		
D_1^0 $\frac{1}{2}(1^+)$	2422.2 ± 1.8	2415.9	4727.7	6.3	34.5	0
$D_2^{*\pm}$ $\frac{1}{2}(2^+)$	2459 ± 4	2451	4796	8	35	0
D_2^{*0} $\frac{1}{2}(2^+)$	2458.9 ± 2.0			8		

Particle $I^G(J^{PC})$		Experimental Mass = M_x	Mass Level = M_l (Mev)	Mass Level (electron mass)	ΔM (Mev) = $M_x - M_l$	n	a
C = S = ± 1							
$D_S^{\pm}$	$0(0^-)*$	1968.6 ± 0.6	1975.0	3865.0	-6.4	28	1
$D_S^{*\pm}$	$0(?^?)$	2112.4 ± 0.7	2116.1	4141.1	-3.7	30	1
$D_{S1}^{\pm}$	$0(1^+)$	2535.35 ± 0.39	2539.31	4969.30	-3.96	36	1
$D_{SJ}^{\pm}$	$0(?^?)$	2573.5 ± 1.7	2572.0	5033.3	1.5	37	-1
B = ± 1							
$B^{\pm}$	$\frac{1}{2}(0^-)*$	5279.0 ± 0.5	5283.1	10338.7	<u>-4.1</u>	76	-1
B^0	$\frac{1}{2}(0^-)*$	5279.4 ± 0.5			-3.7		
B^*	$\frac{1}{2}(1^-)$	5325.0 ± 0.6	5325.5	10421.7	-0.5	75.5	1
B = ± 1, S = ∓ 1							
B_S^0	$0(0^-)*$	5369.6 ± 2.4	5360.8	10490.7	8.8	76	1

$c\bar{c}$

Particle $I^G(J^{PC})$	Experimental Mass = M_x	Mass Level = M_l (Mev)	Mass Level (electron mass)	ΔM (Mev) = $M_x - M_l$	n	a
$\eta_c(1S)$ $0^+(0^{-+})$	2979.8 ± 1.8	2976.1	5824.0	3.7	42.5	0
$J/\psi(1S)$ $0^-(1^{--})$	3096.87 ± 0.04	3093.39	6053.60	3.48	44.5	-1
$\chi_{c0}(1P)$ $0^+(0^{++})$	3415.1 ± 0.8	3421.0	6694.7	-5.9	48.5	1
$\chi_{c1}(1P)$ $0^+(1^{++})$	3510.51 ± 0.12	3510.48	6869.82	0.03	50.5	-1
$\chi_{c2}(1P)$ $0^+(2^{++})$	3556.18 ± 0.13	3562.09	6970.82	-5.91	50.5	1
$\psi(2S)$ $0^-(1^{--})$	3685.96 ± 0.09	3684.26	7209.91	1.70	53	-1
ψ $0^-(1^{--})$	3769.9 ± 2.5	3773.7	7384.9	-3.8	53.5	1
ψ $0^-(1^{--})$	4040 ± 10	4032	7890	8	58	-1
ψ $0^-(1^{--})$	4415 ± 6	4412	8633	3	63	0

$b\bar{b}$

Particle $I^G(J^{PC})$	Experimental Mass = M_x	Mass Level = M_l (Mev)	Mass Level (electron mass)	ΔM (Mev) = $M_x - M_l$	n	a
$\Upsilon(1S)$ $0^-(1^{--})$	9460.30 ± 0.26	9453.96	18500.89	6.34	136	-1
$\chi_{b0}(1P)$ $0^+(0^{++})$	9859.9 ± 1.0	9871.0	19317.1	-11.1	142	-1
$\chi_{b1}(1P)$ $0^+(1^{++})$	9892.7 ± 0.6	9905.8	19385.1	-13.1	142.5	-1

APPENDIX B. WHAT INFORMATION DO ELEMENTARY PARTICLE MASSES GIVE?

Particle $I^G(J^{PC})$	Experimental Mass = M_x	Mass Level = M_l (Mev)	Mass Level (electron mass)	ΔM (Mev) = $M_x - M_l$	n	a
$X_{b2}(1P)$ $0^+(2^{++})$	9912.6 ± 0.5	9910.4	19394.1	2.2	140.5	1
$Y(2S)$ $0^-(1^{--})$	10023.26 ± 0.31	10016.17	19601.11	7.09	142	1
$X_{b0}(2P)$ $0^+(0^{++})$	10232.1 ± 0.6	10227.8	20015.2	4.3	145	1
$X_{b1}(2P)$ $0^+(1^{++})$	10255.2 ± 0.5	10253.4	20065.3	1.8	147.5	-1
$X_{b2}(2P)$ $0^+(2^{++})$	10268.5 ± 0.4	10263.1	20084.2	5.4	145.5	1
$Y(3S)$ $0^-(1^{--})$	10355.2 ± 0.5	10357.7	20269.4	-2.5	149	-1
$Y(4S)$ $0^-(1^{--})$	10580.0 ± 3.5	10580.5	20705.4	-0.5	149	1
Y $0^-(1^{--})$	10865 ± 8	10863	21258	2	154	1
Y $0^-(1^{--})$	11019 ± 8	11018	21562	-1	158.5	-1

Particle $I^G(J^{PC})$	Experimental Mass = M_x	Mass Level = M_l (Mev)	Mass Level (electron mass)	ΔM (Mev) = $M_x - M_l$	n	a
BARYONS						
S=0, I=$\frac{1}{2}$						
p $\frac{1}{2}(\frac{1}{2}^+)$*	938.27200 ± 0.00004	938.4444	1836.4859	<u>-0.1724</u>	13.5	-1
n⁰ $\frac{1}{2}(\frac{1}{2}^+)$*	939.56533 ± 0.00004			1.1209		
S=0, I=$\frac{3}{2}$						
Δ masses are approximate						
Δ++ $\frac{3}{2}(\frac{3}{2}^+)$	1231	1234	2416	-3	17.5	1
Δ+ $\frac{3}{2}(\frac{3}{2}^+)$	1232			<u>-2</u>		
Δ0 $\frac{3}{2}(\frac{3}{2}^+)$	1233.5			-0.5		
Δ-	unknown!					
pole positions						
Δ++	1209.6 ± 0.5	1216.5	2380.6	-6.9	17.5	-1
Δ0	1210.75 ± 0.6			-5.75		

APPENDIX B.   WHAT INFORMATION DO ELEMENTARY PARTICLE MASSES GIVE?

Particle $I^G(J^{PC})$	Experimental Mass = M_x	Mass Level = M_l (Mev)	Mass Level (electron mass)	ΔM (Mev) = $M_x - M_l$	n	a
S= -1,I = 0						
Λ 0($\frac{1}{2}^+$)*	1115.683 ± 0.006	1112.228	2176.576	3.455	16	-1
Λ 0($\frac{1}{2}^-$)	1406 ± 4	1411	2761	-5	20	1
Λ 0($\frac{3}{2}^-$)	1519.5 ± 1.0	1516.5	2967.8	3	21.5	1
S= -1,I = 1						
Σ+ 1($\frac{1}{2}^+$)*	1189.37 ± 0.07	1190.43	2329.61	<u>-1.06</u>	17	0
Σ0 1($\frac{1}{2}^+$)*	1192.642 ± 0.024			2.21		
Σ- 1($\frac{1}{2}^+$)*	1197.449 ± 0.030	1199.119	2346.612	7		
				-1.67	17	1
Σ+ 1($\frac{3}{2}^+$)	1382.8 ± 0.4	1375.5	2691.7	7.3	19.5	1
Σ0 1($\frac{3}{2}^+$)	1383.7 ± 1.0	1375.5	2691.7	8.2	19.5	1
			2720.7	-6.6	20	-1
Σ- 1($\frac{3}{2}^+$)	1387.2 ± 0.5	1390.3	2720.7	<u>-3.1</u>	20	-1

Particle $I^G(J^{PC})$	Experimental Mass = M_X	Mass Level = M_l (Mev)	Mass Level (electron mass)	ΔM (Mev) = $M_X - M_l$	n	a
S = -2, I = $\frac{1}{2}$						
Ξ^- $\frac{1}{2}(\frac{1}{2}^+)$*	1321.31 ± 0.13	1320.77	2584.68	0.54	19	-1
Ξ^0 $\frac{1}{2}(\frac{1}{2}^+)$*	1314.83 ± 0.2			-6		
Ξ^- $\frac{1}{2}(\frac{3}{2}^+)$	1535.0 ± 0.6	1540.56	3014.79	-5.6	22	0
Ξ^0 $\frac{1}{2}(\frac{3}{2}^+)$	1531.80 ± 0.32	1529.32	2992.79	2.48	22	-1
Ξ $\frac{1}{2}(?^?)$	1690 ± 10	1693	3313	-3	24	1
Ξ $\frac{1}{2}(\frac{3}{2}^-)$	1823 ± 5	1821	3563	2	26	0
Ξ $\frac{1}{2}(?^?)$	1950 ± 15	1946	3809	4	28	-1
Ξ $\frac{1}{2}(?)$	2025 ± 5	2031	3974	-6	29	0
S = -3, I = 0						
Ω^- $0(\frac{3}{2}^+)$*	1672.45 ± 0.29	1668.35	3264.86	4.1	24	-1
Ω^- $0(?^?)$	2252 ± 9	2257	4417	-5	32	1

C = 1

Particle $I^G(J^{PC})$	Experimental Mass = M_x	Mass Level = M_l (Mev)	Mass Level (electron mass)	ΔM (Mev) = $M_x - M_l$	n	a
Λ_c^+ $0(\frac{1}{2}^+)*$	2284.9 ± 0.6	2292.4	4486.2	-7.5	32.5	1
Λ_c^+ $0(\frac{1}{2}^-)$	2593.9 ± 0.8	2590.9	5070.3	3.0	37	0
Λ_c^+ $0(\frac{3}{2}^-)$	2626.6 ± 0.8	2626.0	5138.9	0.6	37.5	0
Σ_c^{++} $1(\frac{1}{2}^+)$	2452.8 ± 0.6	2450.9	4796.3	1.9	35	0
Σ_c^+ $1(\frac{1}{2}^+)$	2453.6 ± 0.9			2.7		
Σ_c^0 $1(\frac{1}{2}^+)$	2452.2 ± 0.6			1.3		
Σ_c^{++} $1(\frac{3}{2}^+)$	2519.4 ± 1.5	2520.9	4933.3	1.5	36	0
Σ_c^0 $1(\frac{3}{2}^+)$	2517.5 ± 1.4			-3.4		
Ξ_c^+ $\frac{1}{2}(\frac{1}{2}^+)*$	2466.3 ± 1.4	2467.8	4829.3	-1.5	35.5	-1
Ξ_c^0 $\frac{1}{2}(\frac{1}{2}^+)*$	2471.8 ± 1.4	2468.8	4831.3	3	35	1

Particle	$I^G(J^{PC})$	Experimental Mass = M_x	Mass Level = M_l (Mev)	Mass Level (electron mass)	ΔM (Mev) = $M_x - M_l$	n	a
$\Xi_c'^+$	$\frac{1}{2}(\frac{1}{2}^+)$	2574.1 ± 3.3	2574.6	5038.3	-0.5	36.5	1
$\Xi_c'^0$	$\frac{1}{2}(\frac{1}{2}^+)$	2578.8 ± 3.2			4.2		
Ξ_c^+	$\frac{1}{2}(\frac{3}{2}^+)$	2647.4 ± 2.0	2645.1	5176.3	2.3	37.5	1
Ξ_c^0	$\frac{1}{2}(\frac{3}{2}^+)$	2644.5 ± 1.8			-0.6		
Ξ_c^+	$\frac{1}{2}(\frac{3}{2}^-)$	2814.9 ± 1.8	2815.3	5509.5	-0.4	40.5	-1
Ξ_c^0	$\frac{1}{2}(\frac{3}{2}^-)$	2819.0 ± 2.5			3.7		
Ω_c^0	$0(\frac{1}{2}^+)*$	2704 ± 4	2711	5305	-7	39	-1
B = -1							
Λ_b^0	$0(\frac{1}{2}^+)*$	5624 ± 9	5631	11019	-7	81	-1

Note: For $n = 22$, $a = 0$, the mass level is 1540.56 MeV. The mass of the $\Theta(1540)^+$ is 1539.2 ± 1.6 MeV (Particle Data Group (2004), Review of Particle Physics, *Phys. Lett.* **592**, #1-4, 15 July, p. 71).

B.10   Table for isospin mass differences in Mev

particle symbols are used for their masses

π:	$\pi^{\pm} - \pi^0$	$= 139.57018$	$- 134.9766$	$= 4.5936 \pm 0.0001$
ρ:	$\rho^{\pm} - \rho^0$	$= 766.5$	$- 769$	$= -2.5 \quad \pm 2.1$
K:	$K^{\pm} - K^0$	$= 493.677$	$- 497.672$	$= -3.995 \pm 0.05$
K*:	$K^{*\pm} - K^{*0}$	$= 891.66$	$- 896.10$	$= -4.44 \quad \pm 0.5$
K_2^*:	$K_2^{*\pm} - K_2^{*0}$	$= 1425.6$	$- 1432.4$	$= -6.8 \quad \pm 2.8$
D:	$D^{\pm} - D^0$	$= 1869.3$	$- 1864.5$	$= 4.8 \quad \pm 1$
D*:	$D^{*\pm} - D^{*0}$	$= 2010.0$	$- 2006.7$	$= 3.3 \quad \pm 1$
D_2^*:	$D_2^{*\pm} - D_2^{*0}$	$= 2459$	$- 2458.9$	$= 0.1 \quad \pm 6$
B:	$B^{\pm} - B^0$	$= 5279.0$	$- 5279.4$	$= -0.4 \quad \pm 1$

N:	p - n	$= 938.2720$	-939.5653	$= -1.2933 \pm 0.0001$
Δ:	$\Delta^{++} - \Delta^+$	~ 1231	$- 1232$	~ -1
	$\Delta^+ - \Delta^0$	~ 1232	$- 1233.5$	~ -1.5
	$\Delta^0 - \Delta^-$	~ 1233.5	?	$\sim$?
Σ:	$\Sigma^+ - \Sigma^0$	$= 1189.37$	$- 1192.64$	$= -3.27 \quad \pm 0.1$
	$\Sigma^0 - \Sigma^-$	$= 1192.64$	$- 1197.45$	$= -4.81 \quad \pm 0.05$
Σ:	$\Sigma^+ - \Sigma^0$	$= 1382.8$	$- 1383.7$	$= -0.9 \quad \pm 1.4$
	$\Sigma^0 - \Sigma^-$	$= 1383.7$	$- 1387.2$	$= -3.5 \quad \pm 1.5$
Ξ:	$\Xi^- - \Xi^0$	$= 1321.31$	$- 1314.83$	$= 6.48 \quad \pm 0.3$
Ξ:	$\Xi^- - \Xi^0$	$= 1535.0$	$- 1531.8$	$= 3.2 \quad \pm 0.9$
Σ_c:	$\Sigma_c^{++} - \Sigma_c^+$	$= 2452.8$	$- 2453.6$	$= -0.8 \quad \pm 1.5$
	$\Sigma_c^+ - \Sigma_c^0$	$= 2453.6$	$- 2452.2$	$= 1.4 \quad \pm 1.5$
Ξ_c:	$\Xi_c^+ - \Xi_c^0$	$= 2466.3$	$- 2471.8$	$= -5.5 \quad \pm 3$
Σ_c:	$\Sigma_c^{++} - \Sigma_c^0$	$= 2519.4$	$- 2517.5$	$= 1.9 \quad \pm 3$
Ξ_c:	$\Xi_c^{'+} - \Xi_c^{'0}$	$= 2574.1$	$- 2578.8$	$= -4.7 \quad \pm 6.5$
Ξ_c:	$\Xi_c^+ - \Xi_c^0$	$= 2647.4$	$- 2644.5$	$= 2.9 \quad \pm 3.8$
Ξ_c:	$\Xi_c^+ - \Xi_c^0$	$= 2814.9$	$- 2819.0$	$= -5.1 \quad \pm 4.3$

References

Abramowitz, Milton and Irene A. Stegun, eds. (1972), Handbook of Mathematical Functions (New York: Dover Publications).

Ahlfors, Lars V. (1986), Mobius Transformations in R^n Expressed through 2 X 2 Matrices of Clifford Numbers, *Complex Variables* 5, p. 215-224.

Anderson, James L. (1967), Principles of Relativity Physics (New York: Academic Press).

Arfken, George (1970), Mathematical Methods for Physicists, second ed. (New York: Academic Press).

Armstrong, M. A. (1988), Groups and Symmetry (New York: Springer-Verlag).

Bargmann, V. (1947), Irreducible Unitary Representations of the Lorentz Group, *Annals of Mathematics* 48, #3, July, p. 568-640.

Barut, A. O., and Rolf B. Haugen (1972), Theory of Conformally Invariant Mass, *Annals of Physics* 71, p. 519-541.

Barut, A. O., and R. B. Haugen (1973a), Conformally Invariant Massive Spinor Equations, I, *Nuovo Cimento* 18A, #3, 1 Dicembre, p. 495-510.

Barut, A. O., and R. B. Haugen (1973b), Conformally Invariant Massive and Massless Spinor Equations, II, Solutions of the Wave Equations *Nuovo Cimento* 18A, #3, 1 Dicembre, p. 511-531.

Barut, A. O., P. Budinich, J. Niederle, and R. Raczka, (1994), Conformal Space-Times — The Arenas of Physics and Cosmology, *Foundations of Physics* 24, #11, Nov., p. 1461-1494.

Beardon, Alan F. (1995), The Geometry of Discrete Groups (New York: Springer-Verlag).

Becker, F. (1974), Note on the Masses of Leptons and Baryons, *Lett. Nuovo Cimento* 9, #16, 20 Aprile, p. 637-640.

Binegar, B., C. Fronsdal and W. Heidenreich (1983), Conformal QED, *J. Math. Phys.* 24, #12, Dec., p. 2828-2846.

Boerner, Hermann (1963), Representations of Groups (Amsterdam: North-Holland Publishing Co.).

Brackx, F., R. Delanghe and F. Sommen (1982), Clifford analysis (London: Pitman Books Ltd.).

Budinich, P. and Raczka R. (1993), Global Properties of Conformal Flat

Momentum Space and Their Implications, *Foundations of Physics* **23**, #4, April, p. 599-615.

Burn, R. P. (1991), Groups, A Path to Geometry (Cambridge: Cambridge University Press).

Campbell, John Edward (1903), Introductory Treatise on Lie's Theory of Finite Continuous Transformation Groups (Oxford: Clarendon Press).

Carruthers, Peter (1971), Broken Scale Invariance in Particle Physics, *Physics Reports,* **1**, no. 1, p. 1-29.

Cartan, Henri (1995), Elementary Theory of Analytic Functions of One or Several Complex Variables (New York: Dover Publications).

Caticha, Ariel (1998a), Consistency, amplitudes and probabilities in quantum theory, *Phys. Rev.* **A57**, #3, Mar. p. 1572-1582.

Caticha, Ariel (1998b), Consistency and linearity in quantum theory, *Phys. Lett.* **A244**, 13 July, p. 13-17.

Churchill, Ruel V. (1948), Introduction to Complex Variables and Applications (New York: McGraw-Hill Book Co.).

Cnops, J. (1998), Reproducing Kernels and Conformal Mappings in R^n, *Journal of Mathematical Analysis and Applications* **220**, p. 571-584.

Cobb, Donald D. and Edward R. McCliment (1972), Decomposition of the Principle Series of Unitary Irreducible Representations of SU(2,2) Restricted to the Subgroup SU(1,1) $\otimes$ SU(2), *J. Math. Phys.* **13**, #2, Feb. p. 209-214.

Cornwell, J. F. (1984), Group Theory in Physics (London: Academic Press).

de Azcarraga, Jose A. and Jose M. Izquierdo (1998), Lie Groups, Lie Algebras, cohomology and some applications in physics (Cambridge: Cambridge University Press).

Dirac, P. A. M. (1936), Wave Equations in Conformal Space, *Ann. of Math.* **37**, #2, April, p. 429-442.

Elstrod, J., F. Grunewald and J. Mennicke (1987), Vahlen's Group of Clifford Matrices and Spin-Groups, *Math. Z.* **196**, p. 369-390.

Engstrom, H. T. and Max Zorn (1936), The Transformation of Reference Systems in the Page Relativity, *Phys. Rev.* **49**, May 1, p. 701-702.

Esteve, A. and P. G. Sona (1964), Conformal Group in Minkowski Space. Unitary Irreducible Representations, *Nuovo Cimento* **XXXII**, #2, 16 Aprile, p. 473-485.

Fillmore, Jay P. (1977), The Fiteen-Parameter Conformal Group, *International Journal of Theoretical Physics,* **16**, #12, p. 937-963.

Ford, Lester R. (1972), Automorphic Functions (New York: Chelsea Publishing Co.).

Friedrichs, K. O. (1953), Mathematical Aspects of the Quantum Theory of Fields (New York: Interscience Publishers, Inc.).

Fröhlich, H. (1958), Speculations on the Masses of Particles, *Nuc. Phys.* **7**, p. 148-149.

Fröhlich, H. (1960), Space-time reflexions, isobaric spin and the mass ratio of bosons, *Proc. Roy. Soc.* **A257**, 6 Sept., p. 147-164.

Fulton, T., F. Rohrlich and L. Witten (1962a), Conformal Invariance in Physics, *Rev. Mod. Phy.* **34**, #3, July, p. 442-457.

Fulton, T., F. Rohrlich and L. Witten (1962b), Physical Consequences of a Co-ordinate Transformation to a Uniformly Accelerating Frame, *Nuovo Cimento* Ser. X, **26**, p. 652-671.

Garding, L. and A. S. Wightman (1954), Representations of the Anticommutation Relations, *Proc. Nat. Acad. Science.* **40**, p. 617-621; Representations of the Commutation Relations, *Proc. Nat. Acad. Science.* **40**, p. 622-626.

Gel'fand, I. M., R. A. Minlos and Z. Ya. Shapiro (1963), Representations of the Rotation and Lorentz Groups and their Applications (New York: The Macmillan Company).

Gilbert, John E. and Margaret A. M. Murray (1991), Clifford Algebras and Dirac Operators in Harmonic Analysis (Cambridge: Cambridge University Press).

Glanz, James (1998), Cosmic Motion Revealed, *Science* **282**, 18 Dec., p. 2156-2157.

Gross, Leonard (1964), Norm invariance of Mass-Zero Equations under the Conformal Group, *J. Math. Phys.* **5**, #5, May, p. 687-695.

Haantjes, J. (1937), Conformal Representations of an n-dimensional Euclidean space with a non-definite fundamental form on itself, *Nederl. Acad. Wetensch. Proc. (Math)* **40**, p. 700-705.

Hamermesh, Morton (1962), Group Theory and its Application to Physical Problems (Reading MA: Addison-Wesley).

Hill, E. L. (1945), On Accelerated Coordinate Systems in Classical and Relativistic Mechanics, *Phys. Rev.* **67**, #11,12, June 1 and 15, p. 358-363.

Hill, E. L. (1947), On the Kinematics of Uniformly Accelerated Motion and Classical Electromagnetic Theory, *Phys. Rev.* **72**, #2, July 15, p. 143-149.

Hill, E. L. (1951), The Definition of Moving Coordinates Systems in Relativistic Theories, *Phys. Rev.* **84**, #6, Dec. 15, p. 1165-1168.

Ingraham, R. L. (1998), The Angle-Geometry of Spacetime and Classical Charged Particle Motion, *Int. J. Mod. Phys.* **D7**, #4, p. 603-621.

Jackson, John David, (1963), Classical Electrodynamics (New York: John Wiley and Sons, Inc.).

Jaekel, Marc-Thierry and Serge Reynaud (1995), Vacuum fluctuations, accelerated motion and conformal frames, *Quantum Semiclass. Opt.* **7**, p. 499-508.

Jancewicz, Bernard (1988), Multivectors and Clifford Algebra in Electrodynamics (Singapore: World Scientific Publishing Co.).

Jauch, J. M. and F. Rohrlich (1976), The Theory of Photons and Electrons (New York: Springer-Verlag).

Kastrup, H. A. (1966), Gauge Properties of the Minkowski Space, *Phys. Rev.* **150**, #4, 28 Oct., p. 1183-1193.

Ketov, Sergei V. (1997), Conformal Field Theory (Singapore: World Scientific Publishing Co.).

Kim, Y. S. and Marilyn E. Noz (1986), Theory and Applications of the Poincaré Group (Dordrecht: D. Reidel Publishing Co.).

Kleppner, Daniel and Roman Jackiw (2000), One Hundred Years of Quantum Physics, *Science* **289**, 11 August, p. 893-898.

Kupersztych, J. (1976), Is There a Link between Gauge Invariance, Relativistic Invariance and Electron Spin?, *Nuovo Cimento* **31B**, #1, p. 1-11.

Laue, H. (1972), Causality and the Spontaneous Breakdown of Conformal Symmetry, *Nuovo Cimento* **10B**, #1, 11 Luglio, p. 283-290.

Levitt, Leonard S. (1958), An Empirical Formula for the Masses of the Elementary Particles, *Current Science* **27**, #4, April, p. 131.

Lounesto, Pertti (1997), Clifford Algebras and Spinors (Cambridge: Cambridge University Press).

Lounesto, Pertti and Latvamaa, Esko (1980), Conformal transformations and Clifford Algebras, *Proceedings of the American Mathematical Society* **79**, #4, Aug., p. 533-538.

Lyndon, Roger C. (1989), Groups and Geometry, London Mathematical Society, Lectures Notes Series 101 (Cambridge: Cambridge University Press).

Macfadyen, N. W. (1971a), The Reduction O(3,1) ⊃ O(2,1) ⊃ O(1,1), *J. Math. Phys.* **12**, #3, Mar., p. 492-498.

Macfadyen, N. W. (1971b), Conformal Group in a Poincaré Basis. I. Principle Continuous Series, *J. Math. Phys.* **12**, #7, July, p. 1436-1445.

Macfadyen, N. W. (1973a), Conformal Group in a Poincaré Basis. II. Principle Discrete Series, *J. Math. Phys.* **14**, #1, Jan. p. 57-64.

Macfadyen, N. W. (1973b), Conformal Group in a Poincaré Basis. III. Degenerate Series, *J. Math. Phys.* **14**, #5, May, p. 638-646.

MacGregor, Malcolm H. (1978), The Nature of the Elementary Particles (Springer-Verlag, Berlin).

Mack, G. (1977), All Unitary Ray Representations of the Conformal Group SU(2,2) with Positive Energy, *Commun. Math. Phys.* **55**, p. 1-28.

Mack, G. and Abdus Salam (1969), Finite-Component Field Representations of the Conformal Group, *Annals of Physics* **53**, p. 174-202.

Maekawa, Takayoshi (1979a), Linear representations of any dimensional Lorentz group and computation formulas for their matrix elements, *J. Math. Phys.* **20**, #4, April, p. 691-711.

Maekawa, Takayoshi (1979b), Note on the linear representations of any dimensional Lorentz group and their matrix elements, *J. Math. Phys.* **20**, #7, July, p. 1460-1463.

Maekawa, Takayoshi (1980), On the invariant scalar products and the UIR of SO(n,1), *J. Math. Phys.* **21**, #4, April, p. 675-679.

Margenau, Henry and George Moseley Murphy (1955), The Mathematics of Physics and Chemistry (New York: D. Van Nostrand Co., Inc.).

Martin, George E. (1987), Transformation Geometry, An Introduction to Symmetry (New York: Springer-Verlag).

Miller, Willard, Jr. (1968), Lie Theory and Special Functions (New York: Academic Press).

Mirman, R. (1963), Masses of Elementary Particles, *Bull. Am. Phys. Soc.* **8**, #5, p. 434, P2.

Mirman, R. (1995a), Group Theory: An Intuitive Approach (Singapore: World Scientific Publishing Co.).

Mirman, R. (1995b), Group Theoretical Foundations of Quantum Mechanics (Commack, NY: Nova Science Publishers, Inc.).

Mirman, R. (1995c), Massless Representations of the Poincaré Group, electromagnetism, gravitation, quantum mechanics, geometry (Commack, NY: Nova Science Publishers, Inc.).

Mirman, R. (1998), Differing Interactions Require Baryon and Lepton Conservation (physics/9802048).

Mirman, R. (1999), Point Groups, Space Groups, Crystals, Molecules (Singapore: World Scientific Publishing Co.).

Mirman, R. (2001), Quantum Mechanics, Quantum Field Theory: geometry, language, logic (Huntington, NY: Nova Science Publishers, Inc.).

Misner, Charles W., Kip S.Thorne and John Archibald Wheeler (1973), Gravitation (San Francisco: W. H. Freeman and Co.).

Naimark, M. A. (1964), Linear Representations of the Lorentz Group (New York: The Macmillan Company).

Nambu, Y. (1952), An Empirical Mass Spectrum of Elementary Particles, *Prog. of Theor. Phys.* **7**, p. 595-596.

Nehari, Zeev (1975), Conformal Mapping (New York: Dover Publications).

Neumann, Peter M., Gabrielle A. Stoy and Edward C. Thompson (1994), Groups and Geometry (Oxford: Oxford University Press).

Norton, John D. (1993), General covariance and the foundations of general relativity: eight decades of dispute, *Rep. Prog. Phys.* **56**, p. 791-858.

O'Raifeartaigh, Lochlainn and Norbert Straumann (2000), Gauge Theory: Historical origins and some modern developments, *Rev. Mod. Phy.* **71**, #1, Jan., p. 1-23.

Ohanian, Hans and Remo Ruffini (1994), Gravitation and Spacetime (New York: W. W. Norton Company).

Page, Leigh (1936a), A New Relativity. Paper I: Fundamental Principles and Transformations Between Accelerated Systems, *Phys. Rev.* **49**, Feb. 1, p. 254-268.

Page, Leigh (1936b), Comments on Robertson's Interpretation, *Phys. Rev.* **49**, June 15, p. 946.

Page, Leigh and N. I. Adams, Jr. (1936), A New Relativity. Paper II: Transformations of the Electromagnetic Field Between Accelerated Systems and the Force Equation, *Phys. Rev.* **49**, Mar. 15, p. 466-469.

Panofsky, Wolfgang K. H. and Melba Phillips (1955), Classical Electricity and Magnetism (Cambridge, MA: Addison-Wesley Publishing Co.).

Particle Data Group (2000), Review of Particle Physics, *The European Physical Journal* **C15**, #1-4, p. 1-878.

Pease, Robert L. (1970), Diophantine Quantization: Application of the Methods of Algebraic Number Theory to the Theory of Elementary Particles, *Phys. Rev.* **D2**, #6, 15 Sept., p. 1069-1071.

Piirainen, Reijo (1996), Mobius transformations and conformal relativity, *Foundations of Physics* **26**, #2, p. 223-242, p. 1089-1090.

Porteous, Ian R. (1995), Clifford Algebras and the Classical Groups (Cambridge: Cambridge University Press).

Quercigh, Emanuele and Johann Rafelski (2000), A strange quark plasma, *Physics World* **13**, #10, Oct., p. 37-42.

Quigg, Chris (1999), Aesthetic Science, *Scientific American* **280**, #4, April, p. 125-127.

Riesz, Frigyes and Bela Sz.-Nagy (1990), Functional Analysis (New York: Dover Publications).

Rindler, W. (1960), Special Relativity (Edinburgh and London: Oliver and Boyd).

Robertson, H. P. (1936), An Interpretation of Page's "New Relativity", *Phys. Rev.* **49**, May 15, p. 755-760.

Rose, M. E. (1995), Elementary Theory of Angular Momentum (New York: Dover Publications).

Rosen, Joe (1968), The Conformal Group and Causality, *Annals of Physics* **47**, #3, May, p. 468-480.

Rosen, J. (1969), Mass, Momentum and the Conformal Group, *Nuovo Cimento* Ser. X, **62A**, 21 Luglio, p. 648-660.

Ryan, John (1985), Conformal Clifford Manifolds Arising in Clifford Analysis, *Proc. Roy. Irish Academy*, **85A**, #1, p. 1-23.

Sattinger, D. H. and O. L. Weaver (1986), Lie Groups and Algebras with Applications to Physics, Geometry, and Mechanics (New York: Springer-Verlag).

Schensted, Irene Verona (1976), A Course on the Application of Group Theory to Quantum Mechanics (Peaks Island, ME 04108: NEO Press).

Schottenloher, Martin (1997), A Mathematical Introduction to Conformal Field Theory (Berlin: Springer-Verlag).

Schutz, Bernard F. (1993), A First Course in General Relativity (Cambridge: Cambridge University Press).

Schwartzman, Steven (1994), The Words of Mathematics. An Etymological Dictionary of Mathematical Terms Used in English (Washington, DC: The Mathematical Association of America, The Spectrum Series).

Schweber, Silvan S. (1962), An Introduction to Relativistic Quantum Field Theory (New York: Harper and Row).

Schwerdtfeger, Hans (1979), Geometry of Complex Numbers (New York: Dover Publications).

Streater, R. F., and A. S. Wightman (1964), PCT, SPIN AND STATISTICS, AND ALL THAT (New York: W. A. Benjamin, Inc.).

Sudarshan, E. C. G. and N. Mukunda (1983), Classical Dynamics: A Modern Perspective (Malabar, FL: Robert E. Krieger Publishing Co.).

Umezawa, M. (1964), Nambu-Fröhlich Mass Rule and Okubo–Gell-Mann Mass Formulae, *Nuovo Cimento* **XXXIII**, #5, 1 Settembre, p. 1481-1483.

Vaccaro, John A. (1995), Phase operators on Hilbert space, *Phys. Rev.* **A51**, #4, April, p. 3309-3317.

Varshalovich, D. A., A. N. Moskalev, V. K. Khersonskii (1988), Quantum Theory of Angular Momentum: Irreducible Tensors, Spherical Harmonics, Vector Coupling Coefficients, $3nj$ Symbols (Singapore: World Scientific Publishing Co.).

von Baeyer, Hans Christian (2000), Quantum Mechanics at the End of the 20th Century, *Science* **287**, 17 March, p. 1935.

Voss, David (1999), Making the stuff of the big bang, *Science* **285**, 20 Aug., p. 1194-1197.

Weinberg, Steven (1972), Gravitation And Cosmology: Principles and Applications of the General Theory of Relativity (New York: John Wiley and Sons, Inc.).

Weingarten, Don (1973), Complex symmetries of electrodynamics, *Annals of Physics* **76**, #2, April, p. 510-548.

Weiss, Peter (2000), Seeking the Mother of all Matter, *Science News* **158**, #9, August 26, p. 136-138.

Wess, J. (1960), The Conformal Invariance in Quantum Field Theory, *Nuovo Cimento* **XVIII**, #6, 16 Dicembre, p. 1086-1107.

Wightman, A. S. and S. S. Schweber (1955), Configuration Space Methods In Relativistic Quantum Field Theory. I, *Phys. Rev.* **98**, #3, May 1, p. 812-837.

Yaglom, I. M. (1968), Complex Numbers in Geometry, (New York: Academic Press).

Yale, Paul B. (1988), Geometry and Symmetry (New York: Dover Publications).

Yam, Philip (1997), Exploiting zero-point energy, *Scientific American* **277**, #6, Dec., p. 82-85.

Yao, Tsu (1967), Unitary Irreducible Representations of SU(2,2). I, *J. Math. Phys.* **8**, #10, Oct., p. 1931-1954.

Yao, Tsu (1968), Unitary Irreducible Representations of SU(2,2). II, *J. Math. Phys.* **9**, #10, Oct., p. 1615-1626.

Yao, Tsu (1971), Unitary Irreducible Representations of SU(2,2). III. Reduction with Respect to an Iso-Poincaré Subgroup, *J. Math. Phys.* **12**, #3, Mar., p. 315-342.

Index

Definitions are listed in bold

space-dependent, 43
space-dependent dilatations, 43
space-dependent redefinitions of
    coordinates, 59
space, dual, 29
space, Hilbert, 176
space is rotationally invariant, 54
space-like interval, 53
space, Minkowski, 43, **98**, 120
space of definition, 135
space, real, **179**
space, spin, 179
space too small, 164
spaces, non-positive-definite met-
    ric, **15**
spaces, two, 179
spacetime foam, 173
special conformal transformation,
    8, 108
special curves, 19, **22**
special functions, 106, 165, 244
spectral lines, 47
spectroscopy, 246
spectrum, 6, 142
spectrum-generating algebra, 142
spectrum-generating algebras, 199
speed of light, **12**, 21, 26
sphere, 18
sphere, mapping of the unit, 109
sphere, reflection in the unit, 101
spherical harmonics, 155, 179
spin, 38
spin, intrinsic, 196
spin, invariance of, 37
spin is a form of angular momen-
    tum, 197
spin, magnitude of, 37
spin space, 179
spinning top, 197
spinors, 233
stable objects, 170
standard mass, arbitrary, 9
state, conjugate vacuum, **172**
state-labeling problem, 117
state of a multiparticle system, 184
state, physical, 175
statefunction, **2**
statefunction, accelerated, 30
statefunctions for constant accel-
    eration, 34

statefunction for two charged par-
    ticles, 193
statefunction, gravitational, **210**
statefunctions, phase of the, 60
states for the conformal group, ba-
    sis, 50
states of different numbers of ob-
    jects, 187
states, internal, 179
Stegun, Irene A., 240
stela-, 108
stick, measuring, 46
stimulating further research, 242
Stone's theorem, 163
Stoy, Gabrielle A. 67, 68, 108
straight line, 5
straight line, circle is a, **89**
straight lines, 17, 49
straight lines are intrinsic proper-
    ties, 5
strange properties, 55
Straumann, Norbert, 64
Streater, R. F., 110, 173
stretched, 93
stretched, axis is, 94
strong coupling constant, 208
SU(1,1), 154
su(1,1) algebra, 120
su(2), 142
SU(2) cover of SO(3), 135
SU(2,2), 120, 137
SU(2,2) cover of SO(4,2), 137
SU(3), 198
su(3), 142
su(3,1), 142
SU(6), 199
subalgebra, Cartan **116**, 120
subduced, 243
subgroup is not able to label, 243
subgroup, maximal compact, 223
subgroup, Poincaré, 107
subgroups, decomposition into,
    107
Sudarshan, E. C. G., 194
sum of the angles of a triangle, 45
supergroup, 170
superstition, 42
surface, flat, 55
surface, hyperboloidal, 35
surface of constant phase, 178

0-595-33692-2